GRAIN LEGUMES

Edited by **Aakash Kumar Goyal**

Grain Legumes

http://dx.doi.org/10.5772/61382
Edited by Aakash Kumar Goyal

Contributors

Dil Thavarajah, Jose Luis Chavez-Servia, Elena Heredia-García, Netzahualcoyotl Mayek-Pérez, Elia N. Aquino-Bolaños, Sanjuana Hernández Delgado, José C. Carrillo-Rodríguez, Homar R. Gill-Langarica, Araceli M. Vera-Guzmán, Mourad Latati, Mouhamed Lazali, Samia Benlahrech, Fatima Zohra Hamdani, Hafnaoui Elalia, Riadh Takouachet, Ghania Ounane, Sidi Mohamed Ounane, Ghiles Kaci, Michael Nickerson, Ashish Singhal, Asli Can Karaca, Robert Tyler, Gerd Bobe, Thushanthi Perera, Yumie Takata, Ma. Del Socorro Lopez Cortez, Kangfu Yu

CBS, Edition **2017**

Published by InTech

Janeza Trdine 9, 51000 Rijeka, Croatia

Publishing Process Manager Edi Lipovic

Technical Editor SPi Global

Cover Designer InTech Design Team

Additional hard copies can be obtained from orders@intechopen.com

Grain Legumes, Edited by Aakash Kumar Goyal
p. cm.
Print ISBN 978-953-51-2720-8
Online ISBN 978-953-51-2721-5

INTECH

open science | open minds

Contents

Preface

Grain legumes are a main source of nitrogen-rich edible seeds, providing a wide variety of protein-rich products and constituting a major source of dietary protein in the diets of human population especially for vegetarian diet. Legumes comprise the third largest family of flowering plants and provide important sources of food, fodder, oil, and fiber products. Legume seeds typically contain 20 to 25% protein and are also a rich source of dietary fiber. Legumes such as groundnut and soybean are also major sources of edible oil and other industrial by-products. The ability to fix atmospheric nitrogen makes legumes excellent components within the various farming systems because they provide residual nitrogen and reduce the needs for mineral nitrogen fertilizers by associated nonlegumes. Intensification of low-input agricultural production has led to a rapid increase in soil degradation and nutrient depletion in many parts of the world, constituting serious threats to food production and food security. About 70% more food is needed to feed the growing population. Legumes can play an important role to feed the world as there is lots of scope to increase the production.

This book "Grain Legumes" represents the excellent reviews about ongoing legume research and future prospects. This edited book "Grain Legumes" is an attempt to put together different chapters written by experts in their field. The first chapter discusses the diversity of common bean and its importance in breeding, and Chapter two provides details about lentil use as a whole food for biofortification. The third chapter describes the importance of pulse proteins, and the author reviews their structure and functions. The fourth chapter is a review related to antioxidant properties of legumes. The next chapter is related to agronomy and brings out the importance of intercropping between legumes and cereals (Chapter six). The last chapter (Chapter seven) deals with grain legume consumption for its medicinal properties.

First of all, the editor would like to thank all the authors for their outstanding efforts and timely work in producing such fine chapters. I highly appreciate all the reviewers for their time to review the respective chapters. I would also like to thank InTech and Edi Lipović for his clerical assistance, advice, and encouragement during the development of this book. Last but not least heartfelt thanks go to my family and parents for their love, encouragement, and vision that unveiled in me from my earliest years—the desire to thrive on the challenge of always striving to reach the highest mountain in everything I do.

Aakash Kumar Goyal
International Center for Agriculture Research in Dry Areas (ICARDA),
India

Chapter 1

Diversity of Common Bean (*Phaseolus vulgaris* L.) Landraces and the Nutritional Value of their Grains

José Luis Chávez-Servia, Elena Heredia-García,
Netzahualcoyotl Mayek-Pérez,
Elia N. Aquino-Bolaños,
Sanjuana Hernández-Delgado,
José C. Carrillo-Rodríguez,
Homar R. Gill-Langarica and
Araceli M. Vera-Guzmán

Additional information is available at the end of the chapter

http://dx.doi.org/10.5772/63439

Abstract

Grain legumes are considered major sources of dietary proteins, calories, certain minerals and vitamins, and they are the most widely cultivated and consumed crops worldwide. Among them are the common beans, whose major production volumes came from landraces cultivated in traditional farming systems. The objective of this study was to evaluate the phenotypic diversity of a set of common bean landraces from Mexico based on the agromorphological traits and nutritional composition of the grain in the context of traditional farming systems. Different field and laboratory data were collected and complemented with secondary information published in refereed journals and research reports. The results showed that there are significant differences in the morphological and physiological traits of the plant, pod and grain among groups of common bean landraces of different geographic origins, which were associated with different indigenous groups. Similar patterns were observed in the contents of anthocyanins, polyphenols, flavoinds and minerals as well as antioxidant activity. In the evaluated population groups in each region, there are outstanding populations in terms of agromorphological traits and the nutritional value of the grain that can enable a participatory breeding initiative guided by regional objectives. Some populations from Sierra Norte, Oaxaca, presented higher values in Zn and Fe, and populations from Estado de Mexico exhibited high polyphenol and flavonoid values but stable agronomic behaviour.

Keywords: Biochemical characterization, crop evolution, local seed systems, morphological traits, on farm conservation

1. Introduction

The greatest genetic diversity of wild and cultivated beans is distributed throughout the Americas from northern Mexico through Central America and the Andes to northwest Argentina [1]. Domesticated beans are commonly separated into Andean and Mesoamerican genepools [2], but Mexico has been established as the centre of origin, diversification and domestication of the common bean based on archaeological, ethnobotanical, morphological, biochemical, genetic and isoenzyme evidence [3–6]. The distribution pathways of beans into and across Europe were very complex and occurred through several introduction events from the New World combined with direct exchange between European and Mediterranean countries [5].

Currently, the common bean is distributed in Europe, Asia and Africa, where it presents similarities to Andean and Mesoamerican genepools or forms hybrids between both genepools. For example, it was determined that there is a high hybridization frequency in central Europe but low frequencies in Spain and Italy [5]. In Africa, the landraces are frequently grouped into Andean and Mesoamerican genepools with few introgressions among these groups [6], and this pattern of diversity was also detected in China and India [7, 8]. However, the diversity of the common bean has been studied less in Asia and Africa than in Europe and the Americas.

The landrace concept is useful for naming or distinguishing among cultivated varieties through simple traits that are locally adapted to traditional farming systems [9]. In this context, we use the landraces of the common bean as the unit of diversity of the farm or farm-managed population, which farmers select and sow during every crop cycle. In Latin America, the bean landraces contribute 70–90% of the seed planted by farmers for the production of food grain [10], and according to the Asian Pacific Seed Association, the seed saved by farmers accounts for 80–90% of all of the seeds used in Asia [11]. Therefore, it is important to understand the contribution of such landraces to phenotypic and genetic diversity as well as their contribution to on-farm conservation of diversity and traditional diets.

The phenotypic and genetic diversity of common bean landraces is typically evaluated through morphological traits, phaseolin seed proteins, allozymes, the biochemical-nutritional traits of the grain and DNA markers, with the local populations that are preserved on-farm as references, to describe the population structure, to understand diversification processes and biogeographic distributions, and to define strategies for conservation and utilization [5, 6, 7, 12]. Farmers manipulate common bean populations through the use of the traits of the plants or seeds, which influences the population structure as well as grain quality, i.e., chemical composition [13]. Despite the increasing use of DNA-based markers to estimate the genetic diversity of different landrace collections, the evaluation of phenotypic variation is still crucial for determining the adaptation and agronomic potential as well as the breeding and nutritional values of landraces.

South-central Mexico is part of the Mesoamerican region that is considered one of the world's biodiversity hotspots, and 28 ethnic groups are concentrated in this region, including the Otomi, Mazahua, Nahuatl, Popolucas, Zapoteco, Mixteco, Mixe, Amuzgo, Triqui, Mazateco, Chinanteco, Mayas, Chontales, Huaves, Chatino, Cuicateco, Chontal, Tzetzal, Tzotzil, Purepecha, Totonaco, Ocuilteco and Matlazinca among others. Various studies of common bean in this region have indicated that it contains the greatest genetic diversity of *Phaseolus vulgaris* in a biocultural diversity context [14–16], a fact documented by the diversity hotspot designation [17]. According to passport data from Mexican genebanks and genetic diversity studies, the highest *P. vulgaris* genetic diversity is being preserved in the fields of indigenous farmers distributed along the south-central part of the region, which includes the Mesoamerican, Jalisco and Durango races.

A wide variety of nutritional compounds with multiple positive effects for human health are contained in bean seeds with the high contents of protein, fibre, polyphenols, flavonoids, carotenoids, saponins, oligosaccharides, condensed tannins, lectins, trypsin inhibitors and phytic acid considered to be the most important components. Polyphenols, anthocyanins and flavonoids, among other phytochemical compounds, are particularly related with antioxidant biological activities and preventive effects against cardiovascular or chronic degenerative diseases, such as cancer, obesity and diabetes as well as other conditions related to the metabolic syndrome, triglycerides and cholesterol [18–23].

In East Africa, the per capita consumption rate of the common bean is above 40 kg/year [24], and in Mexico, it is 10.38 kg/person a year with an overall food intake of 5.43 g protein/person a day [25, 26]. As reported by Aguirre-Arenas et al. [27], an annual per capita consumption ranging from 9.8 to 25.9 kg has been estimated in four communities from Morelos, Tamaulipas, San Luis Potosi and Michoacan, Mexico. Additionally, the highest bean production (86.4%) comes from marginal agrosystems with lower fertility, unirrigated soils on slopes where the landraces are usually planted by small-scale farmers [28].

Interest in the use of grain legumes and their constituents in food is growing in many developed countries, and the factors contributing to this trend include access, legumes are cultivated in almost all climatic conditions, as well as their reported nutritional and health benefits. Despite changes in consumer preferences, pulses have a long history of use as human food in many developing countries from the Mediterranean region, Africa, Latin America and Asia, and in some cases, the demand exceeds national production volumes. Peas, chickpeas, lentils, beans, soybeans, mung beans, faba beans and other grain legumes are important sources of food proteins, amino acids, minerals and bioactive substances (phenolic compounds, lectins, enzyme inhibitors, phytates, oligosaccharides), all of which have functional properties that benefit human health and are modified by processing or physical treatments [29–31]. It is necessary to know the nutritional composition and to test these functional properties of the common bean landraces as well as their contribution to traditional rural diets to increase the consumption volume. The main objective of this research was to evaluate the phenotypic diversity of a set of common bean landraces from Mexico through agromorphological traits and the nutritional composition of the grain in the context of traditional farming systems and on-farm seed selection.

2. Diversity of common bean landraces using agro-morphological traits and the criteria of farmers

Diverse researchers around the world use local populations or samples of common bean landraces as reference sets to study their genetic diversity and population structure [6, 7, 32, 33, 34]. Nevertheless, such on-farm crop genetic diversity is highly dynamic as a result of the agroecological conditions of cultivation, the preferences of farmers in seed selection and the management of seed lots, among other factors, that have important impacts on the population structure and chemical composition of the grain, which change across time [5, 24, 33, 35]. In Latin America, from 70 to 90% of the common bean seed that is planted is produced by farmers [10], and in Asia, from 80 to 90% of all seeds used by farmers came from local supplies instead of seed companies [12]. In this context, famers play an important role in the evolutionary process of common beans under domestication, and it is necessary to understand the reasons and criteria for the on-farm management of the bean landraces. In cases where the study samples came from genebanks, such diversity remained static for years, and the genetic diversity estimators differ from those of places where on-farm population dynamics exist, which are primarily in the centres of genetic diversity.

The current genetic and phenotypic variation in the common bean in Mexico in terms of plant and physiological characters and grain and chemical composition is based on the genetic diversity preserved by pre-Colombian cultures, contemporary farmers and the genepools of related wild species [14, 34, 35]. The variety of the shapes, sizes and colours of the grains in several regions where beans are grown is an example of the still poorly documented genetic diversity in the fields of traditional farmers that is usually characterized by agro-morphological descriptors [36], molecular markers [37], and protein [38], anthocyanin [39] and polyphenol [40] contents among others.

The characters that are most commonly used by farmers to differentiate their landraces include grain colour, colour brightness (shiny black, dirty black, etc.), growth type when planted alone (type I and II) or with corn (growth III and IV), time from planting to the harvest of green beans and grain, and the size, shape and colour of the grain and pod. Other more accurate descriptive characters are sometimes used by farmers aiming to distinguish their landraces that are most commonly related to field and post-harvest behaviour, such as high yield, the quantity of harvested pods, tolerance to biotic and abiotic factors, grain hardness and grain-cooking time, among others. This is all local information related to the knowledge shared by local farmers concerning their landraces.

An agromorphological characterization was required to describe the phenotypic variability in native beans collected from populations from different states and regions in Mexico as well as the phenotypic differences among and within the sources of the different landraces. As revealed by the results, significant differences exist among bean landraces from different geographical origins as well as within each geographical location (**Table 1**). Consequently, despite the phenotypic similarities in colour and seed dimensions, the different common bean populations are subject to patterns of isolation in addition to artificial selection, which leads to divergence in the characteristics of agronomic importance.

Descriptive traits	Mexican states (5)	Populations/states	CV (%)	Min.	Max.	Average
Days to flowering	193.43**	66.4**	5.6	49.3	74.3	63.7
Pod length (cm)	8.72**	4.84**	10.4	9.0	15.4	12.8
Pod width (cm)	0.30**	0.07**	7.9	0.8	1.6	1.14
Grains/pod	11.06**	4.22**	6.4	3.9	8.9	6.5
Number of pods/plant	557.31**	149.7**	17.2	2.7	36.7	19.5
Number of pods/exp. parcel	488499**	115605**	16.7	19.7	900.0	471.5
Wet yield/parcel (kg)	144.82**	38.34**	18.4	0.20	17.2	7.4
Dry yield/parcel (kg)	3.06**	0.83**	18.6	0.03	2.31	1.12
Wet weight/pod (g)	51.88**	37.7**	14.0	3.91	20.81	14.9
Dry weight/pod (g)	1.27**	0.88**	10.3	0.59	3.16	2.3
Wet yield/plant (g)	164898.4**	58479**	19.2	26.7	694.0	299.1
Dry yield/plant (g)	3173.88**	1257.1**	18.8	4.2	88.6	45.7
Wet weight/30 pods (g)	46719.49**	33999**	14.0	117.3	624.3	446.4
Dry yield/ 30 pods (g)	1147.61**	799.2**	10.3	17.3	94.7	68.6

**Significant at $P < 0.01$.

Table 1. Mean squares of the analyses of variance, coefficients of variation (CV), minimums, maximums, and average values of agromorphological traits evaluated in common bean landraces from five Mexican states.

An average of 6.5 grains per pod were quantified among the landraces in Mexico, which exceeds the 3.9 grains per pod estimated in 25 native bean populations from different regions of Italy [41], 4–5 grains per pod from 14 common bean varieties in Turkey [42], and 4.2 grains per pod from different varieties of Andean origin in Asturias, Spain [43]. Such observations indicate that most of the landraces evaluated in this study have the dual purpose of consumption as green pods or as dry grain, depending on the length and number of grains per pod, and they are in high demand in regional markets. The level of demand is very important to farmers because they have opportunity to sell their surplus in local markets after fulfilling their food needs.

Overall, there is great variability in the various morphological traits reported for 49 bean populations from different regions of Mexico, and significant differences were detected in bean populations grouped by their state of origin as well as within each state, such as Oaxaca, Puebla, Tlaxcala, Guerrero and the State of Mexico (**Table 2**). In particular, there are differences in plant vigour and the number of days to flowering (from 49 to 74 days), which influence the timing from planting to first harvest. For instance, the populations with a shorter time from planting to harvest were those from Puebla and Guerrero, and this precocity was observed in all of the Puebla bean populations evaluated by Ramírez-Pérez et al. [44] with flowering intervals ranging from 41 to 57 days. The results showed that one fraction of the genetic and phenotypic diversity of common bean landraces is preserved in every region of Mexico, and this diversity is being increased through agro-morphological and physiological traits such as time to flowering, yield per plant and plant development.

Descriptive traits	**Mexican states**				
	Oaxaca	**Puebla**	**Tlaxcala**	**Guerrero**	**Mexico**
Days to flowering	62.1^{b}	66.0^{a}	60.3^{b}	66.4^{a}	62.6^{b}
Pod length (cm)	$12.7^{a,b}$	12.4^{b}	12.4^{b}	13.6^{a}	12.6^{b}
Pod width (cm)	1.0^{d}	1.1^{c}	1.2^{b}	1.1^{c}	1.3^{a}
Grains/pod	7.2^{a}	6.6^{b}	5.9^{c}	6.6^{b}	5.7^{c}
Number of pods/plant	20.4^{a}	21.0^{a}	14.4^{b}	24.5^{a}	14.7^{b}
Number of pods/experimental parcel	474.1^{b}	$456.6^{b,c}$	$349.0^{c,d}$	653.2^{a}	345.0^{d}
Wet yield/exp. parcel (kg)	7.0^{b}	6.7^{b}	5.3^{b}	10.7^{a}	5.8^{b}
Dry yield/exp. parcel (kg)	1.03^{b}	1.06^{b}	0.85^{b}	1.60^{a}	0.88^{b}
Wet weight/pod (g)	13.1^{c}	$14.3^{b,c}$	$14.8^{a,b}$	16.3^{a}	$15.7^{a,b}$
Dry weight/pod (g)	2.0^{c}	2.2^{b}	$2.4^{a,b}$	2.4^{a}	2.4^{a}
Wet yield/plant (g)	$281.7^{b,c}$	308.1^{b}	212.9^{c}	404.0^{a}	$242.7^{b,c}$
Dry yield/plant (g)	$42.5^{b,c}$	$48.4^{a,b}$	34.9^{c}	60.0^{a}	$37.0^{b,c}$
Wet weight/30 pods (g)	392.6^{c}	$428.9^{b,c}$	$443.5^{a,b}$	488.4^{a}	$472.4^{a,b}$
Dry yield/30 pods (g)	59.0^{c}	67.3^{b}	$71.7^{a,b}$	73.3^{a}	72.7^{a}

In rows, means with similar letters are not significantly different (Tukey's test, $P < 0.05$).

Table 2. Comparison of the means of agromorphological traits evaluated in common bean landraces from different Mexican states.

The bean populations originating in Guerrero had a greater pod length (12.7 cm) than those from Mexico, which had the greatest width (1.3 cm). Hence, the highest numbers of pods per plant were yielded in the Oaxaca, Puebla and Guerrero landraces and ranged from 20.4 to 24.5. It is noteworthy that the highest average quantity of grains per pod was reported in the Oaxaca populations (7.2), from which higher pod and grain yield expectations were derived (**Table 2**). The Tlaxcala, Guerrero and Mexico populations had a statistically higher average weight, both fresh and dry, per pod (> 14.5 g), which yielded higher pod and grain yields (**Table 2**). This means that the Oaxaca bean grains are thinner and smaller than those reported in other states, as classified by Espinosa-Pérez et al. [28] using a collection of native common bean populations from the south-central region of Mexico. The common beans from Tlaxcala and Guerrero have a high potential for use in a breeding programme or for direct consumption and regional cultivation, but the Oaxaca beans can be used as sources of genes due to their resilience in environments with limited soil moisture.

One of the limitations to grain legume performance is the low flower sets in environments with moisture stress in the soils and coldness. In addition to the flower sets being low, approximately 70–80% in the floral phase of the buds, the pods fall prematurely with only a fraction reaching maturity. A decrease in the number of pods per plant and final yield occurs in these cases, which affects the adaptability of a bean population to different agroecological production niches [45].

A principal components (PC) analysis was performed once the population morphological characterization from the different states of Mexico had been completed, and 74.9% of the total phenotypic variance in the bean populations was captured in the first two PC. The traits that described the first component (PC1) were pod number and weight per plant, both fresh and dry, and pod width and the weight of 30 pods for PC2. The spatial distribution of the bean population with the highest pod number and weight per plant is in the upper and lower right quadrant (II and III) in **Figure 1**, corresponding to the landraces from Guerrero, Puebla, and Oaxaca as well as some others from the State of Mexico. The phenotypic divergences among geographic groups, shown in **Figure 1**, confirm the previous results in the context of the biogeographic and cultural manipulation of the traditional farming systems by the farmers. For example, the indigenous groups from Oaxaca have a particular form of cultivation related to rainfall conditions and the sowing depth, among other practices, that differ from the management by the farmers of Puebla and Guerrero.

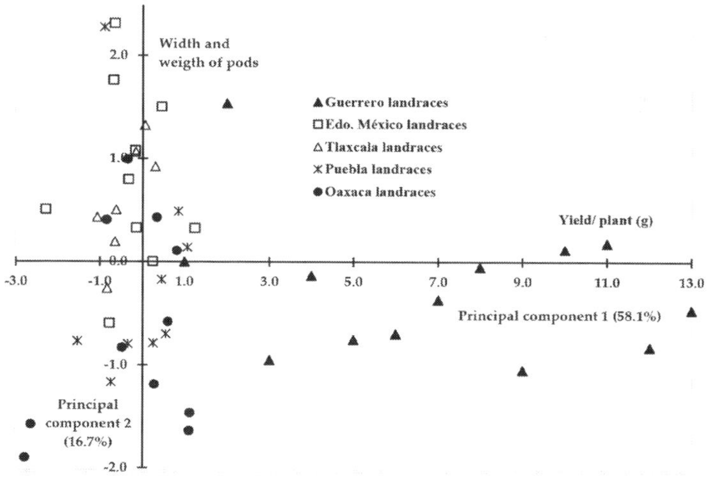

Figure 1. Dispersion of populations of common bean landraces in the states of Oaxaca, Puebla, Tlaxcala, Guerrero and Estado de Mexico based on two principal components of agromorphological traits.

A total of 70 native common bean populations from different geographical regions of Oaxaca, Mexico, which are occupied by the Zapotec, Mixtec, Mixes and Chinantec indigenous groups, were evaluated and compared with 10 improved varieties. Significant differences were detected in the common bean landraces both among and within the geographical regions of origin. Distinctive plant and grain characters were revealed in the bean populations originating from the Mixtec region and cultivated by the Mixtec indigenous group when compared to those grown in the Central Valley (Zapotecs of the Valley) and Sierra Norte (Zapotecs of the Sierra), indicating that differences among native common bean populations are induced by the natural and artificial selection pressures exerted by indigenous groups (**Table 3**). The result highlights the differences in management practices among regions inhabited by indigenous groups that are conferred to their common bean landraces because the agroecological conditions are different in each region.

Descriptive traits	Regions of Oaxaca (5)	Populations/ regions	CV (%)	Min.	Max.	Average
Days to flowering	15247.1**	609.3**	7.4	38.2	101.0	79.6
Pod length (cm)	143.1**	11.70**	6.8	9.8	17.5	13.8
Grains/pod	52.79**	7.85**	10.3	3.4	8.9	6.8
Dry weight of 60 pods (g)	0.040**	0.004**	17.1	40.0	240.0	136.2
Dry weight of grains/60 pods (g)	0.024**	0.002**	18.2	30.0	170.0	101.0
No. of pods/experimental parcel	215499.7**	37456.5**	33.2	44.5	502.7	235.4

**Significant at P < 0.01.

Table 3. Mean squares of the analyses of variance, coefficients of variation (CV), minimums, maximums, and average values of agromorphological traits evaluated in common bean landraces from different regions of Oaxaca, Mexico.

Descriptive traits	Landraces of				Improved varieties
	Sierra Sur	Sierra Norte	Valles Centrales	Mixteca	
Days to flowering	$92.7^{a,1}$	87.8^{b}	$84.7^{b,c}$	81.7^{c}	44.3^{d}
Pod length (cm)	15.5^{a}	12.9^{c}	15.2^{a}	14.0^{b}	10.9^{d}
Grains/pod	7.8^{a}	6.1^{c}	7.7^{a}	6.9^{b}	5.1^{d}
Dry weight of 60 pods (g)	145.9^{a}	149.5^{a}	138.9^{a}	144.7^{a}	77.5^{b}
Dry weight of grains/60 pods (g)	109.1^{a}	106.5^{a}	107.6^{a}	107.8^{a}	55.1^{b}
No. of pods/experimental parcel	279.0^{a}	234.9^{a}	278.4^{a}	247.3^{a}	104.3^{b}

In rows, means with the same letter are not significantly different (Tukey's test, P < 0.05).

Table 4. Comparison of the means of agromorphological traits among common bean landraces from different geographic origins in Oaxaca, Mexico.

It is noteworthy that the bean populations from different regions of Oaxaca were ranked significantly higher in terms of several agronomic and morphological characteristics in comparison with 10 improved varieties used by commercial producers (**Table 4**). The bean populations from the Sierra Sur were late to flower, but they have a similar pattern to that of the bean populations from the Central Valley in relation to grain number per pod and pod length, with averages of 7.8 and 15.5, respectively. These values are higher than those estimated in 15 bean populations from different regions of Jalisco and Nayarit, as reported by Lépiz et al. [35], and moreover, they exceeded the average calculated for 21 common bean genotypes from Tabasco of 4.2 grains per pod [37]. Therefore, the quality of the bean populations from Oaxaca significantly exceeded that of the improved varieties, which means that there is high variability in their agronomic traits, so these populations may be useful as raw materials for a breeding programme.

Additionally, a PC analysis was also carried out to evaluate the overall variability, and in this case, 81.2% of the total variation was captured in the first two PC (**Figure 2**). The descriptive variables of the first component were days to flowering and dry grain weight, and for the second component, they were grain number per pod and average dry weight of 60 pods. Hence, in addition to there being phenotypic differences among bean populations from different states, significant divergences are also denoted among bean populations located in different geographic regions within the same state (such as Oaxaca). All of the local bean populations represent a feasible strategy for bean planting and harvesting by small farmers who plant less than 3 ha in the south-central and south-eastern regions of Mexico. Regionally, the zone of origin of each common bean landrace determines its adaptability; subtropical and tropical row materials have difficulty adapting and producing grain in temperate regions and vice versa.

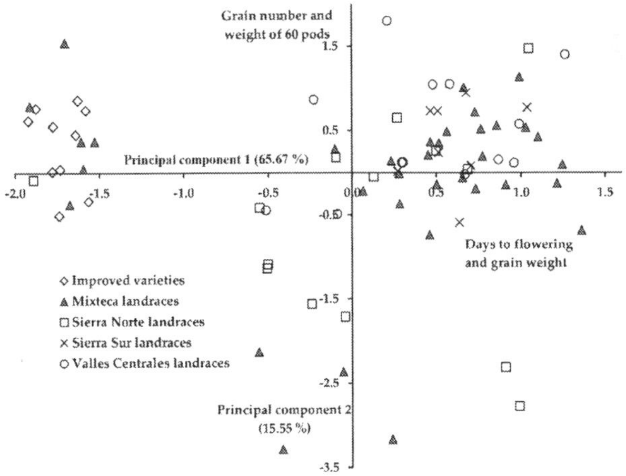

Figure 2. Dispersion of common bean landraces and improved varieties from different regions of Oaxaca, Mexico, based on two principal components of agromorphological traits.

3. Grain nutritional composition in common bean landraces

In terms of the chemical composition of the common bean, it is an outstanding protein source with a low carbohydrate level. Thus, approximately 15 % of the required daily protein intake for a 70-kg adult [46] is provided by a 100-g daily portion of beans. The amino acid content differs from one genotype to another, and it also depends on the ecological conditions for planting, farm management and grain storage conditions, among other factors [23, 47].

The proximal analysis of the common bean indicates that grains contain 14–33% protein, 1.5–6.2% lipids, 14–19% total fibre (from 10.1 to 13.4% insoluble and from 3.1 to 7.6% soluble), 2.9–4.5% ash and 52–76% carbohydrate [48]. Derived and non-derivative (dietary fibre) polysaccharides plus a variety of mono-, di- and polysaccharides are among the carbohydrates that occur in greater proportions. Thus, the grain contains a variety of low and non-digestible carbohydrates, but their functional structure changes through soaking and cooking, increasing the amount of soluble fibre and the digestibility [49, 50].

As assessed by cooking time, there is high variability in protein content and grain hardness among improved varieties and landraces. The protein content in native beans from Hidalgo, Mexico, ranged from 16.0 to 26.9%, as reported by Muñoz-Velázquez et al. [38], with variations in cooking time from 43 to 81 minutes for wine- and creamy yellow-coloured beans, and higher protein content plus a 95% in vitro digestibility rate was observed in light brown Canario and Flor de Mayo varieties. Protein contents ranging from 16.3 to 29.2% with cooking times from 50 to 141 min were reported by Ramírez-Pérez et al. [44] in local, brown-coloured bean populations from Puebla, and protein levels ranging from 21.0 to 25.8% with cooking times from 54 to 118 minutes were reported in local bean varieties from Guerrero by Solano-Cervantes et al. [51]. Certain variations are induced by agroecological or grain management conditions, but such changes are not significant. A constant high protein content through cultivation cycles and years is a characteristic of outstanding genotypes [52].

Regarding essential amino acid content, it has been reported previously that the limiting amino acids in corn grain are apparently complemented by those contained in beans. Phenylalanine, isoleucine, leucine, lysine, methionine, cysteine, threonine, tryptophan and valine are among the main essential amino acids in beans with a range from an average of 1.2 to 1.5 g methionine/100 g of protein and 4.9 to 9.9 g cysteine/100 g of protein. Most amino acids in the grain are found in sufficient quantities to meet the daily requirements of 1.1–6.6 g/100 g of protein [48], and it is noteworthy that the amounts of isoleucine, leucine, lysine, methionine, phenylalanine, threonine and valine contained in the grain do not change significantly after cooking [49].

A rather important and underestimated input provided by human consumption of beans is the portion of minerals, and several authors have reported that the environment has little influence on the differences from genotype to genotype. Instead, differences likely correspond to genetic diversity among and within improved varieties, either wild or cultivated, and landraces [53–55]. The intake of both macro- and micro-minerals is associated with the prevention of prostate cancer [56], and beneficial effects against colon cancer have also been found experimentally in Sprague-Dawley rats [57]. There are other beneficial effects for human health [58], yet the inhabitants of rural and urban communities are deficient in Fe and Zn, which are elements that are mainly associated with malnutrition in children and pregnant women [51]. The mineral content in the common bean varies depending on the genetic material, crop management and grain storage conditions [53, 54, 59].

Significant differences with regards to the S, P, Na, K, Mg, Ca, Fe, Zn, Cu and Mn contents among groups of native bean populations of different geographical origins were recently determined in studies carried out by the authors. The contents were evaluated by means of atomic absorption spectrometry and UV-vis using germplasm from Oaxaca that was planted in an experimental plot, and differences among bean populations from the same geographic region were also determined (**Table 5**). Low S, Na, Ca and Zn contents were presented by the populations originating from the Mixtec region as opposed to the populations from Sierra Norte that had higher S, K, Fe, Zn and Mn contents, which in turn differed from those that originated in the Central Valley with high levels of P, Na, Zn and Cu (**Table 5**). Hence, a relevant fraction of the Mexican *P. vulgaris* genetic pool is in the Oaxaca regions in Mexico, so the genetic pools of the different Oaxaca regions differ in the contents of both mineral macro- and micro-elements. Therefore, the data suggest that common beans provide an important fraction of essential minerals and not only proteins and carbohydrates, and this information is relevant to consumers because the specialized and organic markets demand products with major contents of these minor dietary components.

Sources of variation	**Groups**	**Populations/ groups**	**Error**	**Coef. var. (%)**	**Groups (contents in mg/100 g)**			
					Mixteca	**Sierra Norte**	**Sierra Sur**	**Valles Centrales**
Macro-elements								
S	8094.6**	667.2**	11.7	7.5	39.4[c1]	67.1[a]	41.9[b]	40.5[bc]
P	106769.3**	52431.2**	137.1	3.6	341.7[b]	266.0[c]	267.5[c]	359.8[a]
Na	4327.4**	1017.9**	52.3	10.1	63.9[c]	74.2[b]	70.6[b]	85.1[a]
K	73606.6**	26151.9**	1017.4	3.5	918.4[b]	946.6[a]	909.0[b]	846.4[c]
Mg	746.9**	730.5**	3.8	1.6	117.7[b]	118.6[b]	125.9[a]	113.7[c]
Ca	998.3**	1353.7**	2.8	7.2	91.3[d]	98.3[b]	93.6[c]	100.1[a]
Micro-elements								
Fe	1.83**	2.22**	0.1	6.4	5.24[a]	5.11[a]	5.11[a]	4.87[b]
Zn	4.11**	0.90**	0.17	9.7	4.1[b]	4.5[a]	4.0[b]	4.7[a]
Cu	6.89**	0.97**	0.01	7.4	1.23[b]	1.14[c]	1.25[b]	2.02[a]
Mn	0.17**	0.10**	0.001	2.9	1.24[b]	1.32[a]	1.17[c]	1.18[c]

**Significant at P < 0.01. [1] In rows, means with the same letter are not significantly different (Tukey's test, P < 0.05).

Table 5. Significance of the mean squares of the analyses of variance and comparison of means among groups of common bean landraces from Oaxaca, Mexico, in relation to the mineral contents in the grain.

Significant differences among collections from the groups of different origin were determined by a canonical discriminant analysis (Pillai's trace F = 3.36, and Wilks' lambda F = 3.52; $P < 0.01$). The collection dispersion in reference to the first two canonical discriminant functions is shown in **Figure 3**, and the patterns of differences by geographic origin indicate divergences. For instance, the populations from the Mixtec region are dispersed in the lower left quadrant, very close to those of the Sierra Sur; those from the Sierra Norte are in the lower right quadrant, and the Central Valley has a higher dispersion in all the quadrants. It is also relevant that the samples with high Fe, Cu, Ca, P, S, Mn and K contents exist in the upper right quadrant (**Figure 3**). As a result, the outstanding samples with high mineral contents might be used in a breeding scheme as proposed by Welch and Graham [60], Welch et al. [61] and Teixeira et al. [62] in *P. vulgaris* germplasm. Therefore, as suggested by these authors, more than high yields and adaptability ought to be the main criteria for bean selection.

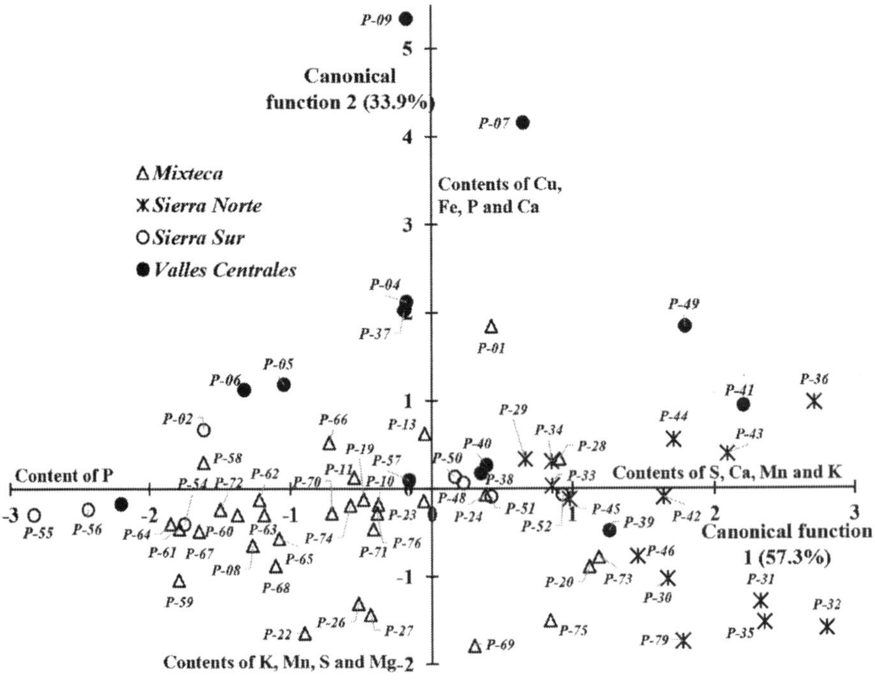

Figure 3. Scatterplot of common bean landraces from different regions of Oaxaca based on two principal canonical functions and the mineral contents in the grain.

The dispersion of the evaluated populations based in the amount of total macro- and micro-element content is shown in **Figure 4**. A total of four scenarios for the populations of particular interest were generated by the creation of four quadrants based on the average content of both macro- and micronutrients. For instance, populations with a higher microelement content are scattered in the upper left quadrant, but these were low in macronutrients. On the other hand, populations that are scattered in the lower right were high in macro- but low in micro-elements. The outstanding populations with higher averages of both macro- and micro-elements are located in the right upper quadrant, where populations from the Mixtec, Sierra Norte and Central Valleys appear. Specifically, the P-06 population is characterized by a high content of both micro- and macro-elements, whereas the P-60, P-67, P-75 and P-79 contain a higher amount of only macro-elements. Consequently, we believe that a set of native bean populations with high macro- and micro-element contents can be identified in every region of Mexico, and they are preserved by farmers and used directly as food (**Figure 4**).

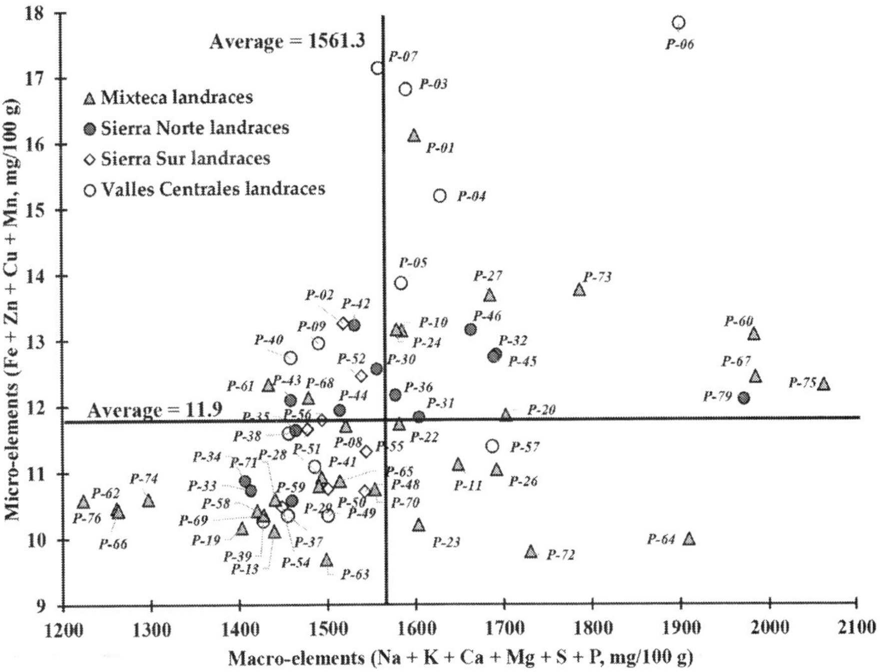

Figure 4. Dispersion of common bean landraces from different regions of Oaxaca in relationship to the total contents of the macro- and micro-elements in the grain.

Another relevant aspect of the bean is its functional compounds and potential nutraceutical content, so 25 native bean populations were collected from Oaxaca, Mexico and experimentally cultivated. At harvest time, a sample ranging from 200 to 500 individuals per population was taken and analysed in a laboratory for the contents of monomeric anthocyanins, polyphenols and flavonoids as well as antioxidant activity by DPPH and a colour index (**Table 6**).

Accesion[1]	An[2]	Seed coat			Seed				Seed colour	
		Po	Fl	AA	Po	Fl	AA	L*	Chrome	h°
EDOMEX-011-11[p]	0.04	82.2	20.9	564.5	2.5	0.70	13.8	59.5	8.0	46.2
EDOMEX-011-7[mc]	0.04	94.6	9.3	751.3	2.5	0.37	14.1	53.5	17.1	58.6
EM-01-01[pr]	1.13	89.5	12.3	667.9	2.7	0.40	11.8	54.4	10.2	50.7
Average-Mexico	*0.4[c2]*	*88.7[c]*	*14.2[b]*	*661.2[c]*	*2.6[a]*	*0.49[a]*	*13.2[b]*	*55.8[b]*	*11.7[ab]*	*51.8[a]*
GRO-01-103[pr]	0.41	67.9	12.8	621.7	2.1	0.33	12.0	51.6	9.4	33.7
GRO-011-15[c]	0.05	61.2	6.5	132.5	2.3	0.30	13.8	68.2	10.7	65.3
GRO-011-16[r]	0.25	55.7	16.8	610.1	2.6	0.78	13.5	50.5	15.5	51.7
GRO-01-118[pr]	0.25	92.5	12.9	724.4	1.3	0.33	10.4	36.9	5.1	21.5
GRO-011-19[pr]	0.34	51.0	12.8	512.9	2.7	0.77	11.0	56.7	8.9	47.0
GRO-011-20[r]	0.07	57.9	14.1	520.7	1.6	0.18	7.1	50.0	9.7	42.5
GRO-011-23[p]	0.22	51.2	19.6	639.4	1.3	0.63	16.6	59.7	9.5	40.8
GRO-10-120[p]	0.53	70.2	15.0	631.4	2.3	0.32	13.1	48.1	11.4	43.6
GRO-10-129[r]	0.42	80.9	8.5	735.9	2.5	0.48	10.9	53.4	7.6	35.2
GRO-10-87[r]	0.59	62.1	21.5	779.2	1.4	0.61	10.2	49.1	12.4	40.3
GRO-10-99[r]	0.24	52.2	11.6	459.7	2.3	0.42	9.1	60.7	6.3	36.4
Average-Guerrero[3]	*0.3[d]*	*63.9[d]*	*13.8[c]*	*578.9[d]*	*2.0[c]*	*0.47[b]*	*11.6[c]*	*53.2[b]*	*9.7[ab]*	*41.6[b]*
OAX-011-07[pr]	0.38	87.9	15.7	750.8	2.7	0.56	8.3	47.8	11.6	44.9
OAX-011-12[y]	0.37	71.4	10.0	615.7	3.3	0.54	20.3	47.7	13.1	48.7
OAX-011-28[b]	2.14	57.0	5.9	534.7	1.9	0.38	10.5	51.7	5.6	67.7
OAX-011-29[mc]	1.54	108.2	7.3	1021.7	2.3	0.30	15.9	63.6	7.7	61.2
OAX-011-30[b]	3.47	127.0	11.0	973.8	1.9	0.35	12.5	49.1	4.8	81.3
Average-Oaxaca	*1.6[b]*	*90.3[b]*	*9.9[e]*	*779.3[b]*	*2.4[b]*	*0.43[c]*	*13.5[b]*	*52.0[b]*	*8.6[b]*	*60.7[a]*
PUE-011-13[p]	0.25	104.5	15.7	713.4	2.0	0.27	11.2	48.5	12.1	35.7
PUE-011-14[y]	0.04	39.2	8.9	389.4	1.3	0.10	7.4	62.1	25.4	72.1
PUE-011-15[cp]	9.07	27.7	7.9	321.6	5.4	0.64	32.4	59.5	11.1	62.0
PUE-011-20[b]	1.94	31.3	8.0	240.4	1.3	0.28	23.0	45.5	4.8	79.4
PUE-011-34[mc]	1.32	80.2	11.1	728.0	1.5	0.32	25.3	56.3	9.2	64.1
PUE-11-33	0.05	70.6	9.8	603.6	2.7	0.49	15.9	53.7	10.5	41.4
Average-Puebla	*2.1[a]*	*58.9[e]*	*10.2[d]*	*499.4[e]*	*2.4[b]*	*0.35[e]*	*19.2[a]*	*54.3[b]*	*12.2[a]*	*59.1[a]*

Accesion[1]	An[2]	Seed coat			Seed			Seed colour		
		Po	Fl	AA	Po	Fl	AA	L*	Chrome	h°
TLA-10-5[c] (average-Tlaxcala)	*0.2*^{e}	*123.4*^{a}	*14.8*^{a}	*985.3*^{a}	*2.6*^{a}	*0.41*^{d}	*13.4*^{b}	*62.2*^{a}	*9.8*^{ab}	*60.6*^{a}
DHS-Tukey	0.02	1.03	0.21	5.9	0.10	0.02	0.65	5.01	3.0	7.52

[1]Origin of groups: EDOMEX/EM= Estado de México; GRO = Guerrero; OAX = Oaxaca; PUE = Puebla; TLA = Tlaxcala.
[2]An = anthocyanins (mg of Cynidin-3-Glucoside-C3G-/g DW); Po = polyphenols (mg gallic acid equivalents–GAE-/g DW); Fl = flavonoids (mg catequine equivalents -CE-/g DW); AA = antioxidant activity (μmol ETrolox/g DW).
[3]among groups, means with the same letter are not significantly different (Tukey's test, $P < 0.05$). Visual colour of the grains: [p] = pink;[mc]= mixture of seed colours; [pr] = pink-red; [c] = cream; [cp] = cream pink; [r] = red; [b] = black; [y] = yellow.

Table 6. Average values of anthocyanins, polyphenols, antioxidant activity and colour index in a Mexican collection of common bean landraces.

The variation among populations in the monomeric anthocyanin content in grains ranged from 0.04 to 9.07 mg cianidine-3-glucoside (C3G)/g on a dry basis, and among groups, the highest average was presented by samples from Puebla (2.1 mg C3G/g) followed by the Oaxaca group (1.6 mg C3G/g) and Tlaxcala (0.2 mg C3G/g). The variation in anthocyanins among the study populations was slightly greater than that described by Gola-Masum-Akond et al. [63] in 29 common bean genotypes of different colours: 0.05 to 0.45 mg C3G/g. Although no specific determinations of anthocyanin types were carried out in this study, it was evident that collections of intense black and red as well as multicoloured beans (a grain mixture of different colours) presented a higher anthocyanin content (>1 mg C3G/g); these included EM-01-01, OAX-011-28, OAX-011-29, OAX-011-30, PUE-011-15, PUE-011-20 and PUE-011-34. As determined by Tsuda et al. [64], delphinidin-3-O-β-D-glucoside, petunidin-O-β-D-glucoside and malvidin-3-O-β-D-glucoside were mostly associated with black bean anthocyanins. However, as reported by Xu et al. [65], the dephinidin-3-glucoside and petunidin-glucoside were the compounds most commonly related to the black grain bean. The highest anthocyanin content in red grain beans was of pelargonidin 3-glucoside, as reported by Choung et al. [66], and higher anthocyanin content was reported in brown, black spotted and pinto grain beans (0.45 a 0.59 mg C3G/g) by Dzomba et al. [67]. As a result, the highest anthocyanin contents in the common bean are associated with beans with a dark seed coat with brown, red and black grain variations.

The anthocyanin contents in black beans reported in this study (1.94 to 3.47 mg C3G/g) were higher than those reported by Salinas-Moreno et al. [39] in 15 black bean varieties, which varied from 0.38 to 0.72 mg C3G/g. Variations are partly attributed to the types of laboratory procedures used, yet differences among genotypes can not be ignored. This was confirmed by the evaluation performed by Xu and Chang [68], who did not find any anthocyanins in the pinto variety but identified delphinidin-3-glucose, malvidin-3, 5-diglucose, petunidin-3-glucose, malvidin-3-galactoside, and malvidin-3-glucose in the black variety (Turtle Eclipse).

Regarding polyphenol and flavonoid contents, important differences were found among the types of seed coats and seeds without seed coat, and the first was favoured in both cases. Among population groups, polyphenol content varied from 58.9 to 123.4 mg GAE/g and from

2.0 to 2.6 mg GAE/g, respectively, plus the variation in flavonoids among groups differed from 9.9 (Puebla) to 14.8 (Tlaxcala) mg CE/g and from 0.35 to 0.49 mg CE/g. These higher polyphenol and flavonoid concentrations provided major antioxidant activity in the seed coat (499.4 to 985.3 μmol ETrolox/g) than in the seed (11.6 a 19.2 μmol ETrolox/g), **Table 5**.

These patterns of high antioxidant activity were repeatedly found among populations within groups such as the PUE-011-15 (27.7 μmol ETrolox/g) and PUE-011-20 (31.3 μmol ETrolox/g) collections, which had the lowest polyphenol contents in the seed coat. Nevertheless, they were higher than the highest seed contents of 20.3 and 32.4 μmol ETrolox/g in the OAX-011-12 and PUE-011-15 collections, respectively (**Table 5**).

Regarding flavonoid contents among populations, the variation ranged from 5.9 (OAX-011-28) to 21.5 (GRO-10-87) mg CE/g in the seed coat and from 0.1 (PUE-011-14) to 0.77 (GRO-011-19) mg CE/g in the seed (**Table 5**). Consequently, the greatest flavonoid content and highest antioxidant activity in the seed coat made us conclude that, with regard to nutraceutical properties, attention should be focused on this fraction, as well as on the grain covering, because of its high potential.

The variation in the total polyphenol content in the grain ranged from 1.3 to 5.4 mg GAE/g in this work, which was slightly lower than that reported by Golam-Masum-Akond et al. [63] in different bean varieties (from 5.9 to 14.1 mg GAE/g) and even lower that the contents in the seed coat (from 27.7 to 127.0 mg GAE/g). These latter values are similar to those reported by Espinosa-Alonso et al. [40] in different common bean populations in Mexico, which ranged from 49.6 to 131 mg GAE/g. Differences in the laboratory methodology could have influenced the results, but populations with potential nutraceutical value due to high flavonoid content in both the seed coat and seed were still detected, such as the OAX-011-29 and TLA-105 among others.

A variation in total flavonoid content ranging from 0.82 to 10.6 mg CE/g in 62 bean populations, both wild and cultivated, was reported by Espinosa-Alonso et al. [40]. The lowest values were similar to those determined for grain but, in several cases, were higher than those revealed for the seed coat for a group of 15 populations, up to 11 mg CE/g. An estimated variation in flavonoids ranging from 0.6 to 1.2 mg CE/g was reported by Boateng et al. [69]. Regardless of the differences in methodology and the estimate derived from the results, the populations under evaluation have important flavonoid levels in the seed coat compared to other genotypes, both cultivated and wild. This fact indicates that the farmers in the study area have a deep knowledge of their bean seeds and continue cultivating the most valued landraces.

The use of combined grain colour indexes (L*: chromaticity, and h°: tone) helps to differentiate populations by the colour of the seed coat or any other characteristic that can be visually appreciated, such as the luminosity index (L*). These indexes became rather useful in the present study for distinguishing the visual colours: pink, cream, yellow or red from other visual variants. Those denominated red and those denominated black were distinguished by the Chroma index, but the tone hue index (h°) was the most accurate because a gradient value was assigned to each colour or variant. Thus, all of the evaluated bean populations were classified in a quantitative way, and in this case, the lowest values correspond to perceptions of pink or red and the highest to black (**Table 5**). These indexes can be used as physical parameters to differentiate between local bean varieties of different grain colour.

The antioxidant activity in the seed coat (132.5 a 1021.7 μmol ETrolox/g) was considerably higher compared to that reported in the seed (7.1 a 32.4 μmol ETrolox/g). It was significantly correlated ($r > 0.36$, $P < 0.05$) with total polyphenol content in the seed coat and seed as well as to the anthocyanins in the grain. As a result, the differentiation among bean groups of origin and bean populations was rather clear. As confirmed by the results of this study, anthocyanins and polyphenols confer high antioxidant activity to the bean grain and seed coat; similar results were reported by Golam-Masum-Akond et al. [63], Dzomba et al. [67], and Oomah et al. [70].

4. Contribution of farmers to on-farm conservation of common bean germplasm

The on-farm conservation of common bean landraces by Mexican small farmers is a basic survival strategy aimed at meeting the daily feeding requirements of rural families. As a consequence, the strategic conservation *in situ* landraces within indigenous, non-indigenous and marginalized communities becomes a way to access food that is not discussed but only conducted to grow and produce beans to eat. However, when there are surpluses, they are sold through either local or regional markets [15, 16, 71]. In several cases, landraces are only regionally or sporadically known nationwide [28] even when remarkable potential has been fully identified in local genetic pools through agronomic, molecular and biochemical assessments [36, 72, 73].

The cultivated wild species *Phaseolus* sp., *Zea mays* ssp. *parviglumis* H. H. Iltis & Doebley (teosinte) and *Cucurbita* sp. [74–77] are also distributed in the Mexican region within Mesoamerica. Possibilities for crossing or genetic flow are generated by the spatial convergence of the genetic pools of wild and cultivated species despite some degree of geographical isolation, differences in flowering time or low crossing rates (<1%). This occurs in beans [78], even though crosses are sometimes high (20–70%) when large numbers of pollinators prevail in the agroecosystems [79].

Beans are grown under different agroecological conditions and for multiple purposes, as we documented in different visits though several regions in the south and southeast of Mexico. The cultivation variants depend on growth rate, both fresh and dry harvest purposes and the levels of precocity. For instance, bean population types III and IV of indeterminate growth, which are most commonly referred to as either climbers or 'frijol de guia', are usually associated with corn and harvested as fresh green beans or dry grain. In these cases, the bean climbs and tangles itself in the corn plant, which being a late flowering and fruiting plant supports the bean. Determinate growth bean types I and II are grown in small plots or backyard gardens to harvest in green beans, and a pink, purple, green with mottled burgundy or simply green colourations characterize the pods, depending on the landraces sold in the local and regional markets. Dry cultivation is performed, and higher yields per unit area are also obtained in such cases.

The bean populations referred to as bush bean plants or determinate growth type I and II are preferred for monoculture planting, most frequently in large areas and in northern Mexico. They are precocious, and the populations display more uniform flowering and fruiting. The grain colour is uniform solid to mottled and variegated, pale white, pink purple, marbled, cream, red, wine, brown, grey, black, white, as well as mottled in different combinations, and it is not surprising to find farmers planting different physical seed mixtures in terms of colour and species. *P. vulgaris* with *Phaseolus coccineus* and *P. vulgaris* with *Phaseolus lunatus* are among the most productive mixtures. All of these observations are consistent with the management of *Phaseolus* sp. diversity described by Worthington et al. [80] and Soleri et al. [81] in Oaxaca, Mexico.

The local bean supply system differs from region to region and from one community to another, and it was revealed through field trips that beans are planted in larger areas by the farmers in the north-central region (>3 ha/producer) than in the south-southeast region (<3 ha/producer) of Mexico. As a consequence, the seed requirement for improved varieties or landraces in both volume and diversity are different in such cases. Improved varieties, and sometimes landraces, are most commonly used by north-central farmers, and often in contrast, landraces, and sometimes improved varieties or even a mixture of both, are most commonly used by the farmers located in the south-southeast. It is a rather common practice for farmers to turn to other communities or regions to obtain seed in years when losses occur due to weather events, such as droughts, storms, floods or hurricanes, or even buy improved varieties. However, they are always looking to find germplasm that suits their agroecological niches [82].

Estimates have been made concerning the movement of seeds within communities in Oaxaca, Mexico, and it was revealed that over 90% of farmers either keep and cultivate their own landrace seed or obtain it from their neighbours or farmers in nearby communities or from traditional local markets [80].

More than a single local bean population has been planted in each agricultural cycle by farmers in Yucatan and some other states in the south-eastern region of Mexico. Some of the reasons underlying the decisions regarding which bean landraces to grow include growth type (I, II, II or IV) because it is directly related with the number of management practices that need to be performed (e.g., Type IV requires more practices); days from planting to the harvest of green beans or dry grain pods; the adaptability to the ecological conditions of the plots or backyard of the producer; tolerance to soil water deficits or low temperature; consumption of fresh forms (as green beans) or dry forms, flowers and/or dry pods; tolerance to insects during storage; grain hardness or consistency with regard to cooking time and flavour and the related organoleptic characteristics, among others [83]. It is appropriate to note that such seed exchange systems are not closed because new seed lots always arrive in the communities being sown, but the sowing continuity of such batches relies on both adaptability and productivity levels in the new places where they are used.

The local seed beans from small farmers are often stored in closed plastic containers and packages, and occasionally, the seed is treated with calcium hydroxide (lime), ash, dried epazote plants (*Dysphania ambrosioides* (L.) Mosyakin et Clemants), chilli pods (*Capsicum* sp.) and chemical insecticides. Additionally, the bean grain is handled differently depending on the harvest volume and the need for storage in the medium term; when only small amounts are harvested (<100 kg), it is generally used for immediate consumption. As a consequence, bags are used and placed in dry spots that are regularly used in the kitchen, but when the harvest is good (>100 kg), the surplus is usually sold at either local or regional markets or even stored in plastic containers with capacities of 100–200 kg that must be perfectly closed. The necessary seed treatments are applied in such cases.

Frequently, farmers from a given region in Mexico or a community have apparently similar bean populations because the beans are alike in grain coat colour, size, shape, growth type, flower and even the local name, as when the Spanish names, such as "negro delgado", "frijol de milpa" or "frijol de cerro", are used or when the local names are used, such as "daá yel-la", "daá laá", "daá tupií" and "daá ya-áá" [81] in the Zapoteca de la Sierra language; "xcolibu'ul" and "tzamá" in the Mayan language [83]; "ndutji" in Mixtec; "etl", "iztac etl", "yahoetl", "pitzahuaqetl" or "itza acaletl" in Nahuatl; "tatsuniutul" in Purepecha; "tsjúú" in Mazahua; "chenek" in Tzotzil; "m'jnai" in Chinanteco; "rune" in Triqui [84], among other indigenous languages. This means that even if the beans are visually or morphologically equal or identified by the same landrace name, they cannot be assumed to be from similar populations. Additionally, the landraces in the *Phaseolus* regions and communities of geographical origin cannot be assumed to have low levels of genetic diversity, mainly in the region known as Lerma-Santiago where a high genetic diversity prevails, based on the documented genetic profiles, geographical origin, phylogeny and ethnohistory of the local bean populations.

The south-southeast regions of Mexico are recognized as the centres of the origin, domestication and genetic diversity of the common bean [4, 77, 85] and where, even today, in indigenous communities, knowledge of the germplasm, crop and seed management [16] is transferred from parents to their children. As a result, the management of genetic diversity in the hands of farmers has established a certain group of features in each bean population that is adapted to each particular agro-ecological niche that is influenced by consumer preferences [85]. Some evidence of such facts was confirmed by the analysis of genetic diversity among different seed samples from farmers in the region of Santa Maria Jaltianguis in the state of Oaxaca, México, where, using SSRs and RFPs markers, significant differences (F_{ST}) were revealed among farmers with similar bean seed lots of the Mesoamerican and Jalisco races [80, 81].

The bean genepools in Mexico can be classified into four groups: a) a total of 85 improved varieties that are currently registered in the Catalogo Nacional de Variedades Vegetales del Sistema Nacional de Inspección y Certificación de Semillas [86] for marketing to farmers; b) approximately 7000 Mexican *P. vulgaris* accessions that are mainly preserved in the germplasm banks from the Instituto Nacional de Investigaciones Forestales, Agricolas y Pecuarias (INIFAP), Universidad Autónoma Chapingo (UACH) and Universidad de Guadalajara; c) out of the 70 *Phaseolus* wild and cultivated species that are distributed in Mexico, the Mexican germplasm banks have the seed of 28 wild species that are distributed throughout the country from sea level to an altitude of 3000 m (**Table 7**). It should be considered that even though both wild and cultivated species exist, five species have already been domesticated in the region including *P. vulgaris, P. coccineus, P. lunatus, P. acutifolius* and *P. dumosus* [35, 87]. d) Finally, the genepools composed of wild species, landraces and heirlooms in the hands of farmers from different Mexican regions; a single farmer may usually hold from 1 to 3 landraces [81, 83]. Now, taking into account that there were 609,342 small family bean production units in 2008 [88], it can be estimated that there are currently 609,342 to 1,828,026 seed lots with a certain degree of differentiation induced by the handling that each farmer provides to his bean populations. As a consequence, each seed lot is designated as the unit of physical diversity that is shaped by all of the bean grains used by each farmer for the next crop, which is treated as independently reproducing a particular type of bean [89, 90]. The highest *P. vulgaris* genetic diversity, which is generally classified as the Mesoamerican, Jalisco and Durango races, is preserved in the fields of small Mexican farmers [2, 4, 14, 77, 91].

Phaseolus species[1]	*Phaseolus* distribution by Mexican state
xanhtotrichus	Hidalgo, Chiapas
vulgaris	Most of the country
tuerckheimii	Chiapas
ritensis	Chihuahua, Durango
polymophus	Aguascalientes, Coahuila, Durango, Guanajuato, Jalisco, Nuevo León
pluriflorus	Durango, México, Jalisco, Michoacán, Morelos, Nayarit, Sinaloa
pedicellatus	Guanajuato, Guerrero, Hidalgo, Jalisco, México, Michoacán, Morelos, Nayarit, Querétaro, San Luis Potosí, Tamaulipas, Veracruz
pauciflorus	Chihuahua, Durango, Guerrero, Jalisco, México, Michoacán, Morelos, Nayarit, Sinaloa, Sonora, Zacatecas
parvulus	Chihuahua, Durango, Nayarit, Sinaloa, Sonora, Zacatecas
oligospermus	Chiapas
oaxacanus	Oaxaca
nesonii	Chiapas, Jalisco, México, Michoacán, Oaxaca, Zacatecas
neglectus	Nuevo León, Tamaulipas
microcarpus	Chiapas, Guanajuato, Durango, Guerrero, Jalisco, Michoacán, Oaxaca, Puebla
micranthus	Jalisco, Michoacán, Nayarit, Sinaloa, Sonora
maculatus	Aguascalientes, Chihuahua, Coahuila, Durango, Guanajuato, Hidalgo, Puebla, Querétaro, San Luis Potosí, Sonora, Tlaxcala, Zacatecas
lunatus	Baja California, Campeche, Chiapas, Colima, Guerrero, Jalisco, México, Michoacán, Morelos, Nayarit, Oaxaca, Sinaloa, Tabasco, Tamaulipas, Veracruz, Yucatán

leptostachyus	Chiapas. Chihuahua, Colima, Durango, Guanajuato, Guerrero, Hidalgo, Jalisco, México, Michoacán, Morelos, Nayarit, Nuevo León, Oaxaca, Puebla, Querétaro, San Luis PotosÍ, Sinaloa, Sonora, Tamaulipas, Veracruz, Zacatecas
jaliscanus	Jalisco, Nayarit, Sinaloa, Michoacán
hintonii	Oaxaca, Morelos, Michoacán
grayanus	Chihuahua, Coahuila, Durango, San Luis PotosÍ, Sonora, Zacateca
glabellus	Chiapas, Hidalgo, Oaxaca, Puebla, San Luis PotosÍ, Tamaulipas, Veracruz
filiformis	Baja California, Chihuahua, Coahuila, Sonora
esperanzae	Hidalgo, México, Michoacán, Puebla
coccineus	Template regions from states of México, Chiapas, Oaxaca, Guerrero, Morelos, Puebla, Veracruz, Tlaxcala, Hidalgo, Guanajuato, Michoacán, Jalisco, Nayarit, Zacatecas, Durango, Nuevo León, Tamaulipas, Sinaloa
chiapasanus	Chiapas, Oaxaca
albescens	Jalisco, Michoacán
acutifolius	Baja California, Chihuahua, Durango, Sonora, Sinaloa. Nayarit, Jalisco, Querétaro, Colima, Coahuila, Guerrero, Michoacán, Oaxaca, Chiapas, Veracruz, Tabasco

[1]Species with seed available at the INIFAP germplasm banks, Universidad Autonoma Chapingo and/or Universidad de Guadalajara.

Table 7. Gene pools of wild species of *Phaseolus* distributed in Mexico [34, 73, 91].

5. Perspectives on the implementation of strategies for the participatory breeding of landraces at the community level

Common bean landraces are an important component of Mexican small-scale farms, and there are numerous landraces that are often highly variable in the plant, physiological, seed, biochemical, genetic and nutritional traits and which usually distinguished by local names or characters. The landraces have particular properties or reputational characteristics for adaptation to local climatic conditions and consumer demand for regional dishes.

The demand for seeds by local farmers depends on the market demand for each improved variety. Improved varieties of grains with light colours are regularly demanded by farmers in the north-central region (i.e., Flor de Mayo, Flor de Junio, Bayo, Cacahuate, Canario, Garbancillo, Mayocoba, Ojo de Cabra, Pinto and some others), and dark-coloured or black varieties (i.e., Jamapa, Negro, etc.) are less likely to be in demand. Small farmers in the south-central-southeastern communities, on the other hand, request a higher number of landraces than improved varieties because they farm plots with a great diversity of agroecological, orographic and altitudinal production niches where improved varieties do not usually thrive. Specific genetic differences have been determined among the seed lots of farmers in the same community [80, 81], who thus provide a high genetic diversity grouped in three common bean races, which are the Mesoamerican, Jalisco and Durango, that are conserved *in situ* by small farmers [2, 4, 78].

Efforts to supply improved bean varieties to farmers who have their products sold in domestic and international markets are being made by INIFAP, research centres and universities. Such farmers can pay from $20 to $100 dls for the amount of seed required to sow a single hectare with a last generation variety or imports. Conversely, small farmers in the south, central and south-eastern regions of Mexico lack economic resources to buy the seeds of improved bean varieties and are more likely to supply themselves with their own seed or to borrow it from their neighbours in either the same community or nearby [81, 82, 92]. Therefore, decentralized plant breeding or a strategy different to the traditional scheme is required to improve bean landraces, which means that breeding programmes need to be either participatory or collaborative to implement *in situ* breeding with the cooperation of breeders and farmers or farmer communities to achieve local and regional objectives in the fields of local farmers. A relevant lack of genetic improvement programmes prevails in Mexico because there is also a lack of bean breeders.

Unique opportunities to use the gene sources of more than 20 wild species distributed throughout Mexico are offered by the many *Phaseolus* landraces and heirlooms and wild or cultivated germplasm genepools, even though the interspecific crossings have not yet been tested. There are also ways to break through the barriers that prevent crosses among species or any other gene transfer strategies of agronomic and biochemical-nutritional relevance among related or different species in terms of genetic divergences. These underutilized or underexplored opportunities require further study. Genetic markers help to both locate and identify specific groups of genes of agronomic and nutritional biochemistry importance, thus making the genetic selection more efficient. However, investment in laboratory infrastructure as well as equipment and human resources is still required to make assisted breeding with genetic markers a reality. Recent improvements include the generation of advanced lines with varying degrees of resistance to pests and diseases through interspecific hybridization [93].

Evidence of the nutritional and nutraceutical potential of landraces as protein sources (essential amino acids), carbohydrates, minerals and polyphenols (anthocyanins, flavonoids, phenols, carotenoids and some other compounds) with high antioxidant activity were previously above. Small farmers, in several cases, take direct advantage of landraces despite little knowledge of the enormous nutritional contribution that comes from the consumption of common bean landraces. The data generated in universities and research centres must be disseminated to consumers because of the decreasing tendency in *per capita* consumption in Mexico from 18.9 kg in 2000 to 10.2 kg/person/year in 2008 [85, 94]. Important progress was achieved by Welch et al. [61], Blair et al. [95] and Gelin et al. [96] with regards to common bean improvement with the selection of elite germplasm with high Zn and Fe contents. However, despite having a quantitative inheritance, such characters interact with the environment and crop management. The most remarkable outcomes were realized using germplasm from the Andes and Mesoamerica.

One challenge for common bean breeding is the generation of improved varieties, which requires the exploitation of genetic variability and the application of local knowledge. At local and regional levels, a farmer is aware that genotypes or landraces respond favourably to abiotic and biotic stresses, including future scenarios of climate change. As previously reported, most breeding for drought resistance has been within the Mesoamerican gene pool and based on grain yield under stress. The sources for drought resistance originated from Jalisco and Durango, Mexico [97]. The impacts of both abiotic and biotic stresses on the common bean crop are influenced by interactions with other environmental components, such as soil texture characteristics, organic matter content, the degree to which aggregate stability affects water infiltration, soil water-holding capacity, and the ability of the roots to acquire moisture and nutrients.

The highest phenotypic and genetic diversity of the landraces is in the custody and preserved in the plots, backyards and homes of small farmers in Mexico. Such self-generated seed producers are able to exchange this diversity among neighbours and relatives who require only small quantities, which is a different scheme than that employed by seed companies or institutions that provide improved seed varieties because the demand in the communities is lower than the minimum required by a business aiming to multiply the improved varieties. Furthermore, the latter are not always adapted to the agroecological niches of small traditional producers, so the local seed exchange systems become the only sources of supply for small farmers, who require different breeding strategies than those used in commercial agriculture.

6. Conclusions

To understand the diversity of common bean landraces, to take advantage of the nutritional value of the grain, and to promote strategies for on-farm conservation and utilization, it is necessary to characterize and evaluate the phenotypic and genetic variation managed by traditional farmers, which provides us with a better understanding of the dynamic and structure of cultivated populations. The farmers modify landrace diversity through management practices in accordance with the diverse reasons or criteria used to satisfy their food needs, the agroecological production conditions, cultural factors and, sometimes, market demands.

The results of this study showed that two patterns of diversity in the common bean landraces can be distinguished in Mexico in terms of the geographic area being represented; at the level of states and regions within a state, the landraces are defined by the agromorphological characteristics and chemical composition of the grain, such as the contents of minerals, flavonoids, polyphenols and antioxidant activity as well as grain colour indexes (L*, chrome and hue). The agroecological conditions of cultivation and farm management influence the high variability in the agromorphological and chemical composition of the grain in the common bean landraces.

In each collection of the evaluated common bean landraces, populations were detected with high agronomic and grain composition potential. For example, there were populations with a high number of grains and yield per plant and/or populations with high contents of micro- and macro-elements, polyphenols, flavonoids and antioxidant activity within each level of diversity represented, the Mexican states and the regions in the state of Oaxaca. Therefore, there is germplasm available at both diversity levels to start a breeding programme at the national level or for on-farm seed selection. In addition, different populations were identified with a dual purpose, the production of both green and dry beans.

In developing countries such as Mexico, consumer preferences are changing towards a decrease in the consumption of common beans, but contradictorily, the incidence of diabetes, obesity and others chronic degenerative diseases is increasing in the population. Therefore, in countries with the major genetic diversity of the species, the common bean is losing its social role. Currently, different researchers are publishing articles demonstrating the protective effect of green or dry beans in the prevention of diverse diseases, including cancer, and other research groups are demonstrating the functional properties. However, there is scarce or no research oriented towards solving the social problem of malnutrition, which is also associated with the reduction in the information available to consumers and non-experts. Today, it is not enough to demonstrate that high genetic diversity exists in common beans with the accompanying nutritional and nutraceutical potential; we must test its utility to solve social problems.

Acknowledgements

The authors are grateful for the financial support provided by the National Polytechnic Institute (project nos. 1636 and 1752); CONACYT-Ciencia Basica (project no. 181756); and COFAA-IPN, EDI-IPN, and S.N.I. fellows.

Author details

José Luis Chávez-Servia[1*], Elena Heredia-García[2], Netzahualcoyotl Mayek-Pérez[3], Elia N. Aquino-Bolaños[4], Sanjuana Hernández-Delgado[3], José C. Carrillo-Rodríguez[5], Homar R. Gill-Langarica[3] and Araceli M. Vera-Guzmán[1]

*Address all correspondence to: jchavezs@ipn.mx

1 National Polytechnic Institute, CIIDIR-Oaxaca, Mexico

2 National Institute of Forest, Agricultural and Livestock Research (INIFAP), Experimental Station-El Bajio, Celaya, Mexico

3 National Polytechnic Institute, Center of Genomic Biotechnology, Reynosa, Mexico

4 Basic Sciences Institute-Universidad Veracruzana, Jalapa, Mexico

5 Technologic Institute of the Oaxaca Valley, Oaxaca, Mexico

References

[1] Singh BD, Gutierrez JA, Molina A, Urrea C, Gepts P. Genetic diversity in cultivated common bean: II. Marker-based analysis of morphological and agronomic traits. Crop Science. 1991a;31:23–29. DOI: 10.2135/cropsci1991.0011183X003100010005x

[2] Singh BD, Gepts P, Debouck DG. Races of common bean (*Phaseolus vulgaris,* Fabaceae). Economic Botany. 1991b;45:379–396. DOI: 10.1007/BF02887079.

[3] Papa R, Gepts P. Asymmetry of gene flow and differential geographical structure of molecular diversity in wild and domesticated common bean (*Phaseolus vulgaris* L.) from Mesoamerica. Theoretical and Applied Genetics. 2003;106:239–250. DOI: 10.1007/s00122-002-1085-z.

[4] Bitocchi E, Nanni L, Bellucci E, Rossi M, Giardini A, Zeuli PS, Logozzo G, Stougaard J, McClean P, Attene G, Papa R. Mesoamerican origin of the common beans (*Phaseolus vulgaris* L.) is revealed by sequence data. In: Proceedings of the National Academy of Sciences of the USA; 2012. 109:E788–E796.

[5] Angioi SA, Rau D, Attene G, Nanni L, Bellucci E, Logozzo G, Negri V, Spagneoletti-Zeuli PL, Papa R. Beans in Europe: origin and structure of the European landraces of *Phaseolus vulgaris* L. Theoretical and Applied Genetics. 2010;121:829–843. DOI: 10.1007/s00122-010-1353-2

[6] Asfaw A, Blair MW, Almekinders C. Genetic diversity and population structure of common bean (*Phaseolus vulgaris* L.) landraces from the East African highlands. Theoretical and Applied Genetics. 2009;120:1–12. DOI: 10.1007/s00122-009-1154-7

[7] Zhang X, Blair MW, Wang S. Genetic diversity of Chinese common bean (*Phaseolus vulgaris* L.) landraces accessed with simple sequence repeat markers. Theoretical and Applied Genetics. 2008;117:629–640. DOI: 10.1007/s00122-008-0807-2.

[8] Kumar V, Sharma S, Sharma AK, Kumar M, Sharma S, Malik S, Singh KP, Sanger RS, Bhat RS. Genetic diversity in indicant common bean (*Phaseolus vulgaris* L.) using random amplified polymorphic DNA markers. Physiology and Molecular Biology of Plants. 2008;14:383–387. DOI:

[9] Camacho TC, Maxted N, Scholten M, Ford-Lloyd B. Defining and identifying crop landraces. Plant Genetic Resources: Characterization and Utilization. 2006;3:373–384. DOI: 10.1079/PGR200591

[10] Arenas WC, Cardozo CI, Baena M. Analysis of seed systems in Latin American countries. Acta Angronomica 2015;64:223–229. DOI: http://dx.doi.org/10.15446/acag.v64n3.43985

[11] GRAIN. New mega-treaty in the pipeline: what doses RCEP means for farmers' seeds in Asia? GRAIN, 07 March 2016. DOI: URL:/e/5405

[12] Blair MW, Soler A, Cortés AJ. Diversification and population structure in common beans (*Phaseolus vulgaris* L.). PLoS One 2012;7:e49488. DOI: 10.1371/journal-pone.0049488

[13] Piergiovanni AR, Lioi L. Intalian common bean landraces: history, genetic diversity and seed quality. Diversity. 2010;2:837–862. DOI: 10.3390/d2060837

[14] Hernández-López VM, Vargas-Vásquez MLP, Muruaga-Martínez JS, Hernández-Delgado S, Mayek-Pérez N. Origin, domestication and diversification of common beans: advances and perspectives. Revista Fitotecnia Mexicana. 2013;36:95–104.

[15] Bermeo A, Couturier S, Galeana PM. Conservation of traditional smallholder cultivation systems in indigenous territories: Mapping land availability for milpa cultivation in the Huasteca Poblana, Mexico. Applied Geography. 2014;53:299–310. DOI: 10.1016/j.apgeog.2014.06.003.

[16] Boege E. The Biocultural Patrimony of the Indigenous Villages from Mexico. Toward *in situ* conservation of the biodiversity and agrobiodiversity in the indigenous territories. Instituto Nacional de Antropología e Historia y Comisión Nacional para el Desarrollo de los Pueblos Indígenas, Mexico, D.F.; 2010. 342 p.

[17] Gorenflo LJ, Romaine S, Mittermeier RA, Walker-Painemilla K. Co-occurrence of linguistic and biological diversity in biodiversity hotspots and high biodiversity wilderness areas. In: Proceedings of National Academy of Science of the USA; 2002. 109:8032–8037.

[18] Leterme P. Recommendations by health organizations for pulse consumption. British Journal of Nutrition. 2002;88:S239–S242.

[19] Leterme P, Muñoz LC. Factors influencing pulse consumption in Latin America. British Journal of Nutrition. 2002;88:S251–S254.

[20] Aparicio-Fernández X, García-Gasca T, Yousef GG, Lila MA, González-de Mejía E, Lorca-Piña G. Chemopreventive activity of polyphenolics from black Jamapa bean (*Phaseolus vulgaris* L.) on HeLa and HaCaT Cells. Journal of Agricultural and Food Chemistry. 2006;54:2116–2122. DOI: 10.1021/jf052974m.

[21] Reynoso-Camacho E, Ramos-Gomez M, Loarca-Pina, G. Bioactive components in common beans (*Phaseolus vulgaris* L.). In: Guevara-González RG, Torres-Pacheco I, editors. Advances in Agricultural and Food Biotechnology. Research Signpost, Kerala, India; 2006. 217–236.

[22] Hernández-Saavedra A, Mendoza-Sánchez M, Hernández-Montiel HL, Guzmán-Maldonado HS, Loarca-Piña GF, Salgado LM, Reynoso-Camacho R. Cooked common beans (*Phaseolus vulgaris*) protect against β-cell damage in streptozotocin-induced diabetic rats. Plant Foods for Human Nutrition. 2013;68:207–212.

[23] Suárez-Martínez SE, Ferriz-Martínez RA, Campos-Vega E, Elton-Punete JE, de la Torre-Carbot K, García-Gasca T. Bean seeds: leading nutraceutical source for human health. CyTA-Journal of Food. 2015: DOI: 10.1080/19476337.2015.1063548.

[24] Blair MW, González LF, Kimani PM, Butare L. Genetic diversity, inter-gene pool introgression and nutritional quality of common bean (*Phaseolus vulgaris* L.) from Central Africa. Theoretical and Applied Genetics. 2010;121:237–248. DOI: 10.1007/s00122-010-1305-x

[25] FAOSTAT. Production by crops and country 2013: Dry beans [Internet]. 2013. Statistics Division, Food and Agriculture Organization of the United Nations. Rome, Italy. Available from: http://faostat3.fao.org/download/Q/QC/E [accessed: 2015-11-28].

[26] Service of Agro-food and Fishing Information (SIAP). Statistics of Agricultural production 2014 in Mexico [Internet]. 2015. Servicio de Información Agroalimentaria y Pesquera (SIAP), Secretaria de Agricultura, Ganadería, Desarrollo Rural, Pesca y Alimentación (SAGARPA). Mexico, D.F. Available from: http://www.siap.gob.mx/cierre-de-la-produccion-agricola-por-cultivo/ [accessed: 2015–11–28].

[27] Aguirre-Arenas J, Escobar-Pérez M, A. Chávez-Villasana A. Evaluation of the food consumption and nutrition in four rural communities. Salud Pública en México 1998;40:398–407.

[28] Espinosa-Pérez EN, Ramírez-Vallejo P, Crosby-Galván MM, Estrada-Gómez JA, Lucas-Florentino B, Chávez-Servia JL. Classification of common dry bean landraces from the south-center of Mexico by seed morphology. Revista Fitotecnia Mexicana. 2015;29:29–38.

[29] Khattab RY, Arnfield SD, Nyachoti CM. Nutritional quality of legume seeds as affected by some physical treatments, part 1: protein quality evaluation. LWT-Food Science and Technology. 2009;42:1107–1112. DOI: 10.1016/j.lwt.2009.02.009

[30] Campos-Vega R, Loarca-Piña G, Oomah BD. Minor components of pulses and their potential impact on human health. Food Research International. 2010;43:461–482. DOI: DOI: 10.1016/j.foodres.2009.09.004

[31] Boye J, Zare F, Pletch A. Pulse proteins: processing characterization, functional properties and applications in food and feed. Food Research International. 2010;43:414–431. DOI: 10.1016/j.foodres.2009.09.003

[32] Móller BSF, Pappas GJ, Valdisser PAMR, Coelho GRC, de Menezes IPP, Abreu AG, Borba TCO, Sakamoto T, Brondani C, Barros EG, Vianello RP. An operational SNP panel integrated to SSR marker for the assessment of genetic diversity and population structure of the common bean. Plant Molecular Biology Reporter. 2015;33:1697–1711. DOI: 10.1007/s11105-015-0866-x

[33] Bitocchi E, Bellucci E, Giardiani A, Rau D, Rodriguez M, Biagetti, Santilocchi R, Spagneoletti-Zeuli PL, Logozzo, Attene G, Nanni L, Papa R. Molecular analysis of the

parallel domestication of the common bean (*Phaseolus vulgaris*) in Mesoamerica and the Andes. New Phytologist. 2013;197:300–313. DOI: 10.1111/j.1469-8137.2012.04377.x

[34] Lépiz R, Ramírez D. The Wild Relatives of Common Beans in the Western of Mexico. Universidad de Guadalajara, México; 2010. 64 p.

[35] Lépiz R, López JJ, Sánchez JJ, Santacruz-Ruvalcaba F, Nuño R, Rodríguez E. Morphological traits of cultivated, wild and weedy forms in climbing common bean. Revista Fitotecnia Mexicana. 2010;33:21–28.

[36] Vargas-Vázquez MLP, Muruaga-Martínez JS, Pérez-Herrera P, Gill-Langarica HR, Esquivel-Esquivel G, Martínez-Damián MA, Rosales-Serna R, Mayek-Pérez N. Morphoagronomic characterization of the INIFAP core collection of the cultivated from common bean. Agrociencia. 2008;42:787–797.

[37] Vidal-Barahona A, Lagunes-Espinoza LC, Valadez E, Ortiz-García CF. Morphological and molecular variability among native and commercial cultivars of black bean in Tabasco, Mexico. Revista Fitotecnia Mexicana. 2006;29:273–281.

[38] Muñoz-Velázquez EE, Rubio-Hernández D, Bernal-Lugo I, Garza-García R, Jacinto-Hernández C. Characterization of native bean genotypes from the state of Hidalgo based on its grain quality. Agricultura Técnica en México. 2009;35:426–435.

[39] Salinas-Moreno Y, Rojas-Herrera L, Sosa-Montes E, Pérez-Herrera P. Anthocyanin composition in black bean (*Phaseolus vulgaris* L.) varieties grown in Mexico. Agrociencia. 2005;39:385–394.

[40] Espinosa-Alonso LG, Lygin A, Widholm JM, Valverde ME, Paredes-López O. Polyphenols in wild and weedy Mexican common beans (*Phaseolus vulgaris* L.). Journal of Agricultural and Food Chemistry. 2006;54:4436–4444. DOI: 10.1021/jf060185e.

[41] Scarano D, Rubio F, Ruiz JJ, Rao R, Corrado G. Morphological and genetic diversity among and within common bean (*Phaseolus vulgaris* L.) landraces from the Campania region (Southern Italy). Scientia Horticulturae. 2014;180:72–78. DOI: 10.1016/j.scienta.2014.10.013.

[42] Ceylan A, Necip Öcal N, Mikail Akbulut M. Genetic diversity among the Turkish common bean cultivars (*Phaseolus vulgaris* L.) as assessed by SRAP, POGP and cpSSR markers. Biochemical Systematics and Ecology. 2014;54:219–229. DOI: 10.1016/j.bse.2014.01.014.

[43] Rivera A, Roselló S, Casañas F. Seed curvature as a useful marker to transfer morphologic, agronomic, chemical and sensory traits from Ganxet common bean (*Phaseolus vulgaris* L.). Scientia Horticulturae. 2015;197:476–482.DOI: 10.1016/j.scienta.2015.10.006.

[44] Ramírez-Pérez AR, Díaz-Ruiz R, Jacinto-Hernández C, Paredes-Sánchez JA, Garza-García R. Diversity of native beans from different regions of the state of Puebla. Revista Mexicana de Ciencias Agrícolas. 2012;3:467–480.

[45] Cruz-Balarezo J, Camarena-Mayta F, Pierre-Baudoin J, Huaringa-Joaquín A, Blas-Sevillano R. Agromorphological evaluation and molecular characterization of ñuña (*Phaseolus vulgaris* L.). IDESIA (Chile), 2010;27:29–40.

[46] Paredes M, Becerra V, Tay J. Inorganic nutritional composition of common beans (*Phaseolus vulgaris* L.) genotypes race Chile. Chilean Journal of Agricultural Research. 2009;69:486–495.

[47] Prolla IRD, Garcia-Barbosa R, Veeck APL, Augusti PR, da Silva LP, Ribeiro ND, Emanuelli T. Cultivar, harvest year and storage conditions affecting nutritional quality of common beans (*Phaseolus vulgaris* L.). Ciência e Tecnologia de Alimentos. 2010;30(Supl. 1): 96–102. DOI: 10.1590/S0101-20612010000500016.

[48] Guzmán SH, Acosta JA, Álvarez-Muñoz MA, García-Delgado S, Loarca-Piña G. Food quality and nutraceutical potential of bean (*Phaseolus vulgaris* L.). Agricultura Técnica en México. 2002;28:159–173.

[49] Barampama Z, Simard RE. Effects of soaking, cooking and fermentation on composition, in-vitro starch digestibility and nutritive value of common beans. Plant Foods for Human Nutrition. 1995;48:349–365. DOI: 10.1007/BF01088494.

[50] Reynoso R, Ríos MC, Torres I, Acosta JA, Palomino AC, Ramos M, González E, Guzmán SH. Common bean (*Phaseolus vulgaris* L.) consumption and its effects on colon cancer in Sprague-Dawley rats. Agricultura Técnica en México. 2007;33:43–52.

[51] Solano-Cervantes F, Díaz-Ruíz R, Jacinto-Hernández J, Aguirre-Álvarez L, Huerta-de la Peña A. Agricultural practices, morphologic, proteinic and culinary description of the grain of beans cultivars sowed in the region of Tlatzala, Guerrero. Ra Ximhai. 2009;5:187–199.

[52] Santalla M, Fueyo MA, Rodino AP, Montero I, de Ron AM. Breeding for culinary and nutritional quality of common bean (*Phaseolus vulgaris* L.) in intercropping systems with maize (*Zea mays* L.). Biotechnology, Agronomy and Society and Environment. 1999;3:225–229.

[53] Moraghan JT, Grafton K. Genetic diversity and mineral composition of common bean seed. Journal of the Science of Food and Agriculture. 2001;81:404–408. DOI: 10.1002/1097-0010(200103)81:4<404::AID-JSFA822>3.0.CO;2-H.

[54] Beebe S, Gonzalez A, Rengifo J. Research on trace minerals in the common bean. Food and Nutrition Bulletin. 2000;21:387–391. DOI: 10.1177/156482650002100408.

[55] Guzmán-Maldonado SH, Acosta-Gallegos J, Paredes-López O. Protein and mineral content of a novel collection of wild and weedy common bean (*Phaseolus vulgaris* L.). Journal of the Science of Food and Agriculture. 2000;80:1874–1881. DOI: 10.1002/1097-0010(200010)80:13<1874::AID-JSFA722>3.0.CO;2-X.

[56] Bidoli E, Talamini R, Bosetti C, Negri E, Maruzzi D, Montella M, Franceschi S, La Vecchia C. Macronutrients, fatty acids, cholesterol and prostate cancer risk. Annals of Oncology. 2005;16:152–157. DOI: 10.1093/annonc/mdi010.

[57] Campos-Vega R, Loarca-Piña G, Oomah BD. Minor components of pulses and their potential impact on human health. Food Research International. 2010;43:461–482. DOI: 10.1016/j.foodres.2009.09.004.

[58] Welch RM. Linkages between trace elements in food crops and human health. In: Alloway BJ, editor. Micronutrients Deficiencies in Global Crop Production. Springer Netherlands, Heidelberg, Germany; 2008. p. 287–309.

[59] Gouveia CSS, Freitas G, de Brito JH, Slaski JJ, Pinheiro-de-Carbalho MAA. Nutritional and mineral variability in 52 accessions of common bean varieties (*Phaseolus vulgaris* L.) from Madeira Island. Agricultural Sciences. 2014;5:317–329. DOI: 10.4236/as.2014.54034.

[60] Welch RM, Graham RD. Breeding for micronutrients in staple food crops from human nutrition perspective. Journal of Experimental Botany. 2004;55:353–364. DOI: 10.1093/jxb/erh064.

[61] Welch RM, House WA, Beebe S, Cheng Z. Genetic selection for enhanced bioavailable levels of iron in bean (*Phaseolus vulgaris* L.) seeds. Journal of Agricultural and Food Chemistry. 2000;48:3576–3580. DOI: 10.1021/jf0000981.

[62] Teixeira RKS, Lima DC, Abreu AFB, Ramalho MAP. Implications of early selection for grain color on iron and zinc content and productivity of common beans. Plant Breeding. 2015;134:193–196. DOI: 10.1111/pbr.12241.

[63] Golam-Masum-Akond ASM, Khandaker L, Berthold J, Gates L, Peters K, Delong H, Hossain K. Anthocyanin, total polyphenols and antioxidant activity of common bean. American Journal of Food Technology. 2011;6: 385–394. DOI: 10.3923/ajft.2011.385.394.

[64] Tsuda T, Watsnabe M, Ohshima K, Norinobu S, Choi S, Kawakishi S, Osawa T. Antioxidative activity of the anthocyanin pigments cyaniding 3-O-β-D-glucoside and cyaniding. Journal of Agricultural and Food Chemistry.1994;42: 2407–2410. DOI: 10.1021/jf00047a009.

[65] Xu BJ, Yuan SH, Chang SKC. Comparative analyses of phenolic composition, antioxidant capacity, and color of cool season legumes and other selected food legumes. Journal of Food Science. 2007;72:S167–S177. DOI: 10.1111/j.1750-3841.2006.00261.x.

[66] Choung M, Choi B, An Y, Chu Y, Cho Y. 2003. Anthocyanin profile of Korean cultivated kidney bean (*Phaseolus vulgaris* L.). Journal of Agricultural and Food Chemistry. 2003;51: 7040–7043. DOI: 10.1021/jf0304021.

[67] Dzomba P, Togarepi E, Mupa M. Anthocyanin content and antioxidant activities of common bean species (*Phaseolus vulgaris* L.) grown in Mashonaland Central, Zim-

babwe. African Journal of Agricultural Research. 2013;8:3330–3333. DOI: 10.5897/AJAR12.225.

[68] Xu B, Chang SKC. Total phenolic, phenolic acid, anthocyanin, flavan-3-ol, and flavonol profiles and antioxidant properties on pinto and black beans (*Phaseolus vulgaris* L.) as affected by thermal processing. Journal of Agricultural and Food Chemistry. 2009;57:4754–4764. DOI: 10.1021/jf900695s.

[69] Boateng J, Verghese M, Walker LT, Ogutu S. Effect of processing on antioxidant contents in selected dry beans (*Phaseolus* spp. L.). LWT-Food Science and Technology. 2008;41:1541–1547. DOI: 10.1016/j.lwt.2007.11.025.

[70] Oomah BD, Cardador-Martínez A, Loarca-Piña G. Phenolics and antioxidative activities in common beans (*Phaseolus vulgaris* L.). Journal of the Science of Food and Agriculture. 2005;85:935–942. DOI: 10.1002/jsfa.2019.

[71] Romero-Arenas O, Damian MA, Rivera JA, Baez A, Huerta M, Cabrera E. The nutritional value of beans (*Phaseolus vulgaris* L.) and its importance for feeding of rural communities in Puebla-Mexico. International Research Journal of Biological Science. 2013;2:59–65.

[72] Rocha-Guzman NE, Gallegos-Infante JA, González-Laredo RF, Cardoza-Cervantes V, Reynoso-Camacho R, Ramos-Gomez M, García-Gasca T, de-Anda-Salazar A. Evaluation of culinary quality and antioxidant capacity for Mexican common beans (*Phaseolus vulgaris* L.) canned in pilot plant. International Food Research Journal. 2013;20:1087–1093.

[73] Hernández-Delgado S, Muruaga-Martínez JS, Vargas-Vázquez MLP, Martínez-Mondragón J, Chávez-Servia JL, Gill-Lagarica HR, Mayek-Pérez N. Advances in genetic diversity analysis of *Phaseolus* in Mexico. In: Caliskan M, Oz GC, Kavakli IH, Ozcan B editors. Molecular Approaches to Genetic Diversity. Vol. 1, InTech, Rijeka, Croacia; 2015. p. 47–73. DOI: 10.5772/60029.

[74] Freytag GF, Debouck DG. Taxonomy, distribution, and ecology of the genus *Phaseolus* (Leguminosae-Papilionoidea) in North America, Mexico, and Central America. Botanical Research Institute of Texas, Forth Worth, TX, USA; 2002. 298 p.

[75] Wilkes G. Corn, strange and marvelous: But is a definitive origin known. In: Smith, editor. Corn: Origin, History, Technology, and Production. John Wiley & Sons, New Jersey, USA; 2004. 3–63.

[76] Montes-Hernandez S, Eguiarte LE. Genetic structure and indirect estimates of gene flow in three taxa of *Cucurbita* (Cucurbitaceae) in western Mexico. American Journal of Botany. 2002;89:1156–1163. DOI: 10.3732/ajb.89.7.1156.

[77] Kwak M, Kami JA, Gepts P. The putative Mesoamerican domestication center of *Phaseolus vulgaris* is located in the Lerma-Santiago basin of Mexico. Crop Science. 2009;49:554–563. DOI: 10.2135/cropsci2008.07.0421.

[78] Ferreira JJ, Alvarez E, Fueyo MA, Roca A, Giraldez R. Determination of the outcrossing rate of *Phaseolus vulgaris* L. using seed protein markers. Euphytica. 2000;113:259–263. DOI: 10.1023/A:1003907130234.

[79] Ibarra-Perez FJ, Ehdaie B, Waines JG. Estimation of outcrossing rate in common bean. Crop Science. 1997;37:60–65. DOI: 10.2135/cropsci1997.0011183X003700010009x.

[80] Worthington M, Soleri D, Aragón-Cuevas F, Gepts P. Genetic composition and spatial distribution of farmer-managed *Phaseolus* bean planting: An example from a village in Oaxaca, Mexico. Crop Science. 2012;52:1721–1735. DOI: 10.2135/cropsci2011.09.0518.

[81] Soleri D, Worthington M, Aragón-Cuevas F, Smith SE, Gepts P. Farmers' varietal identification in a reference sample of local Phaseolus species in the Sierra Juárez, Oaxaca, Mexico. Economic Botany. 2013;67:283–298. DOI: 10.1007/s12231-013-9248-1.

[82] Castiñeras L, Cristóbal R, Pinedo R, Collado L, Arias L. Seed supply networks and limitations from the informal system. In: Hermann M, Amaya K, Latournerie L, Castiñeras L, editors. How farmers preserve their seeds in the humid tropics of Mexico, Cuba, and Peru?, Experiences in a research project on informal seed systems. Bioversity International, Rome, Italy; 2009. p.73–83.

[83] Latournerie L, Yupit EC, Tuxill J, Mendoza M, Arias LM, Castañón G, Chavez-Servia JL. Traditional systems of seed bean and squash storage in Yaxcaba, Yucatan. Revista Fitotecnia Mexicana. 2005;28:47–53.

[84] Brown CH. Prehistoric chronology of the common bean in the new world: The linguistic evidence. In: Staller JE, Carrasco MD, editors. Pre-Columbian Fooways: Interdisciplinary Approaches to Food, and Markets in Ancient Mesoamerica. Springer Science, New York, USA; 2010.273–291. DOI: 10.1007/978-1-4419-0471-3_11.

[85] Angioi SA, Desiderio F, Rau D, Bitocchi E, Attene G, Papa R. Development and use of chloroplast microsatellites in *Phaseolus* spp. and other legumes. Plant Biology 2009;11:598–612. DOI: 10.1111/j.1438-8677.2008.00143.x.

[86] National Service of Inspection and Certification of Seeds (SNCS). National Catalog of Plant Varieties 2014. Secretaria de Agricultura, Ganadería, Desarrollo Rural, Pesca y Alimentación (SAGARPA), Mexico D.F.; 2014. 82 p.

[87] Delgado-Salinas A, Bibler R, Lavin M. Phylogeny of the genus *Phaseolus* (Leguminosae): A recent diversification in an ancient landscape. 2006;31:779–791. DOI: http://dx.doi.org/10.1600/036364406779695960.

[88] González-Estrada A. Estimation of the agrarian and economic structures of corn and bean production in Mexico. Revista Mexicana de Economía Agrícola y de los Recursos Naturales. 2009;2:7–29.

[89] Louette D, Charrier A, Bethaud J. In situ conservation of maize in Mexico: Genetic diversity and maize seed management in a traditional community. Economic Botany. 1997;51:20–38. DOI: 10.1007/BF02910401.

[90] Díaz LM, Blair MW. Race structure within Mesoamerican gene pool of common bean (*Phaseolus vulgaris* L.) as determined by microsatellite markers. Theoretical and Applied Genetics. 2006;114:143–154. DOI: 10.1007/s00122-006-0417-9.

[91] López JL, Ruiz JA, Sánchez JJ, Lépiz R. Climatic adaptation of 25 species of wild bean (*Phaseolus* spp) in Mexico. Revista Fitotecnia Mexicana. 2005;28:221–230.

[92] Latournerie L, Moreno V, Fernández L, Pinedo R, Tun JM, Tuxill J. Traditional system of seed storage: Importance and implications for the agrobiodiversity conservation. In: Hermann M, Amaya K, Latournerie L and L. Castiñeras L, editors. How farmers preserve their seeds in the humid tropics of Mexico, Cuba, and Peru?, Experiences in a research project on informal seed systems. Bioversity International, Rome, Italy; 2009. p. 61–72.

[93] Beaver JS, Osorno JM. Achievements and limitations of contemporary common bean breeding using conventional and molecular approaches. Euphytica. 2009;168:145–175. DOI: 10.1007/s10681-009-9911-x.

[94] Sangerman-Jarquín DM, Acosta-Gallegos JA, Schwenstesius-de-Rindermann R, Damian-Huato MA, Larqué-Saavedra BS. Considerations and social importance of the bean crop in Central Mexico. Revista Mexicana de Ciencias Agrícolas. 2010;1:363–380.

[95] Blair MW, Astudillo C, Beebe S. Analysis of nutritional quality traits in an Andean recombinant inbred line population. Annual Report of Bean Improvement Cooperative. 2005;48:52–53.

[96] Gelin JR, Forster S, Grafton KF, McClean PE, Rojas-Cifuentes GA. Analysis of seed Zn and other minerals in a recombinant inbred population of navy bean (*Phaseolus vulgaris* L.). Crop Science. 2007;47:1361–1366.

[97] Teran H, Singh SP. Comparison of sources and lines selected for drought resistance in common bean. Crop Science. 2002;42:64–70.

Chapter 2

Lentil (*Lens culinaris* Medikus): A Whole Food Rich in Prebiotic Carbohydrates to Combat Global Obesity

Dil Thavarajah , Pushparajah Thavarajah ,
Casey R. Johnson and Shiv Kumar

Additional information is available at the end of the chapter

http://dx.doi.org/10.5772/62567

Abstract

Lentil (*Lens culinaris* Medik.) is a cool season food legume that is high in protein (20–30%) and in a range of micronutrients (e.g., minerals, carotenoids, folates) but very low in phytic acid. Recent research indicates that lentil contains a wide array of low-molecular weight carbohydrates (LMWC) or prebiotic carbohydrates, such as mono- and disaccharides, raffinose-family oligosaccharides (RFO), fructooligosaccharides (FOS), and sugar alcohols, and high-molecular weight resistant starches. Lentil provides more than 13 g of prebiotic carbohydrates per 100 g serving, and this level increases almost two-fold upon cooking, cooling, and reheating. In addition, prebiotic carbohydrate levels vary with lentil genotype and growing location/country. Intestinal microbiome and prebiotic studies suggest a prebiotic-rich, low-calorie diet can reduce the prevalence of obesity and related non-communicable diseases. Lentil thus represents a whole food source of prebiotics that can play a role in efforts to reduce obesity and non-communicable diseases. This chapter provides an overview of the current obesity-related health issues, holistic approaches to reduce obesity, worldwide lentil production, and the promise of pulses, mainly lentil, to be a whole food solution to combat global obesity. In addition, lentil's superior LMWC profile and the genetic potential for further enrichment of prebiotic carbohydrates are briefly discussed.

Keywords: lentils, low-molecular weight carbohydrates, prebiotics, gut microbiome, obesity

1. Introduction

Obesity has become an epidemic. Chronic, non-communicable diseases associated with obesity result in 36 million deaths globally each year, more than all other causes combined [1]. Obesity has been a severally neglected global public health concern for decades [2] and, today, "globesity" —the global epidemic of overweight and obesity—is taking over many parts of the world despite continued economic development [2]. In 1995, 200 million adults were obese; by 2000, 300 million adults were obese; and today, more than 1.9 billion adults are overweight and 600 million are obese [2]. Both obesity and overweight increase the risk of health conditions including hypertension, adverse lipid concentrations, type 2 diabetes, and several cancers (endometrial, breast, prostate, and colon) [3]. This situation calls for immediate public health awareness to reduce obesity and the risk of related health disorders.

Obesity and overweight are preventable health conditions. Several prevention approaches are available, including dietary therapy (low caloric diets), changes in physical activity and/or social behavior, surgical procedures, and combinations thereof. However, because of the nature of these metabolic disorders, solutions will by necessity have a focus on diet. The intestinal microbiome and a prebiotic-rich, low-caloric diet are beginning to be recognized as being important for preventing obesity. Prebiotic-rich diets change microbial species in the human gut, which leads to increased satiety, regulation of intestinal motility, production of short-chain fatty acids, prevention of diarrhea and constipation, and reduction of pathogen colonization [4, 5]. Furthermore, consumption of a prebiotic-rich diet may stimulate the immune system [6], promote mineral absorption (especially iron and selenium), decrease the risk of colon cancer [7, 8], reduce cholesterol levels and excess circulation of glucose, and improve insulin sensitivity [9]. As such, products high in prebiotics are becoming more popular health-promoting foods around the world. Lentil (*Lens culinaris* Medik.), also known as *poor man's meat*, is one such "superfood" and has the potential to provide daily prebiotic requirements. Compared with cereal food products, prebiotics are found at high levels in lentils [10]. The objective of this chapter is to provide an overview of current obesity-related health issues, holistic food approaches to prevent obesity, global lentil production, and the promise of pulse crops, particularly lentil, as a whole food solution to combat global obesity.

2. Obesity

2.1. Global obesity prevalence

Since the 1960s, obesity and overweight prevalence has increased enormously, resulting in a global health burden far surpassing infectious diseases. The numbers tell a clear story. From 1980 to 2013, the number of overweight or obese individuals rose from 857 million to 2.1 billion [11–13]. Worldwide, 37% of men and 38% of women are overweight or obese [13]. In some regions, the rate of obesity in adults exceeds 50% [13]. The prevalence of overweight and obesity among children and adolescents is also rising [11], with these conditions affecting 23.8% of boys and 22.6% of girls in developed countries and 12.9% of boys and 13.4% of girls in developing countries, respectively [13]. Unfortunately, awareness of the problem has not resulted in any decline in obesity and overweight rates, especially in developing countries [13].

Health effects and comorbidities associated with obesity are numerous and include cardiovascular disease, type 2 diabetes mellitus, chronic kidney disease, osteoarthritis, hypertension, stroke, dyslipidemia, gall bladder disease, and some cancers [14–16]. The severity of many of these comorbidities increases with the degree of obesity, i.e., increasing body mass index (BMI) [16]. In an analysis of risk factors contributing to global disease burden, high BMI alone is the sixth highest risk factor for chronic disease even before the consideration of associated comorbidities [14].

The costs of obesity are very high, both for afflicted individuals and for national healthcare systems [14, 17–19]. Obese individuals have impaired physical function and health-related quality of life [19], as well as socioeconomic and emotional consequences such as decreased work force, completion of fewer years of school, stigmatization, decreased self-esteem, and increased likelihood of experiencing poverty [18, 20]. Furthermore, obesity and overweight cause an estimated 3.4 million deaths worldwide, accounting for 4% of life years lost and 4% of disability-adjusted life years [14]. The financial burden of overweight and obesity is equally high, consuming between 0.7 and 9.1% of total healthcare expenditures among countries included in a meta-analysis from 1990 to 2009 [17]. Had associated comorbidities been included, these expenditure estimates would be significantly higher. The etiology of obesity is complex, but three contributing factors have been identified: diet, metabolic dynamics, and physical inactivity [21]. The complexity of obesity arises because of interactions of these factors with genetics and environmental stimuli (e.g., stress). To illustrate, over 150 gene loci relate to obesity and diabetes through effects on body processes, including insulin and insulin receptors, fat deposition and distribution, lipolysis, and hypothalamic function [22]. Obesity is associated with chronic low-grade inflammation, which subsequently leads to a host of downstream pathological conditions [23–26]. Proinflammatory markers, such as interleukin-6 and tumor necrosis factor (TNF)-α and -β, are found in higher concentrations in the liver of obese individuals, resulting in local and systemic insulin resistance [27].

2.2. What causes obesity?

Popkin et al [12]. suggest an evolutionary rationale for the obesity pandemic, pointing out clashes between human biology and the technological and industrial revolutions (**Table 1**). Three problems the authors identify relate directly to diet: sweet preference, disconnect between thirst and hunger mechanisms, and fatty food preference. Food processing, added sweeteners, caloric beverages, and ease of vegetable oil production exploited these biological tendencies, resulting in dramatically increased consumption of sugars, oils, and milled grains. The per capita supply of fats increased from 47 to 82 g/capita/day and of sugars increased from 47 to 61 kg/capita/day between 1961 and 2010; however, the per capita supply of pulse crops (including lentil) dropped significantly over the same period (**Figure 1**) [28]. Furthermore, food availability per capita overall has significantly increased [11].

Biology	Technology
Sweet preferences	Cheap caloric sweeteners, food processing benefits
Thirst and hunger/satiety mechanisms not linked	Caloric beverage revolution
Fatty food preference	Edible oil revolution, high-yield oilseeds, cheap removal of oils
Desire to eliminate exertion	Technology in all phases of movement/exertion
Reproduced from ref [12].	

Table 1. Technological clashes with human biology.

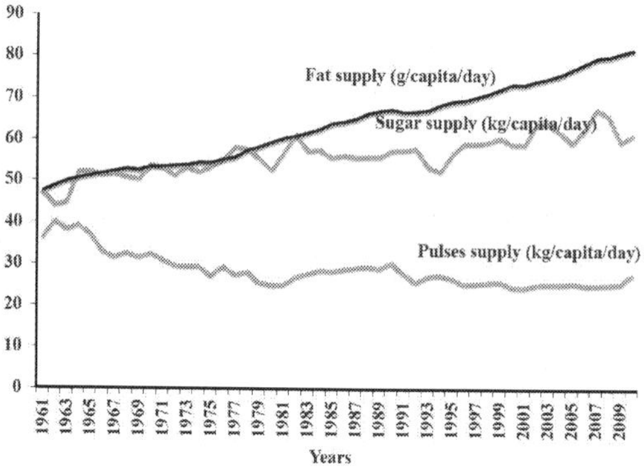

Figure 1. Global supply of sugar, fat, and pulses from 1961 to 2009 [28].

2.3. Holistic approaches to preventing obesity

Consumption of foods rich in sugars/fat and low physical activity are major contributors to obesity. Environment and human genetics also play a fair role, but societal changes are driving the obesity epidemic. In 2002, experts from United Nations, Food and Agricultural Organization (FAO), and World Health Organization (WHO) collaborating centers joined forces to identify an immediate action plan to prevent global obesity. They recommended (1) correcting social food intake and physical activity patterns, (2) intervention and commitment to the above actions (food intake and physical activity) at all levels (e.g., individual through to community, national, and international levels), and (3) developing new government policies for populations to improve individual lifestyle characteristics [29].

Economic growth, urbanization, and globalization of food markets have led to nutritional transitions around the world. Consequently, consumption of high-fat high-energy diets and/or reduced physical activity have become common practice. For thousands of years, world agriculture adopted simple rotations of profitable cash crops, i.e., wheat, maize, and rice, with nutritionally rich legume crops. However, modern calorie-centric agriculture is devoid of these traditional or more diverse cropping systems, leading to nutrition transitions that are linked

http://dx.doi.org/10.5772/62567

to increased rates of obesity in both developing and developed countries. To overcome these nutritional transitions, appropriate food systems should feature a host of activities related to the distribution, utilization, and consumption of nutritious foods including fruits, vegetables, cereals, and pulse crops [30]. Holistic systems approaches use healthy soil, water, seeds, fertilizer, human labor, and capital and have outputs beyond food—primarily the long-term health of both the people and the environment. Such approaches have proven successful with respect to the prevention of micronutrient malnutrition in developing countries. For example, introduction of biofortified staple food crops through HarvestPlus to severely malnourished populations has promoted distribution, utilization, and consumption of these food crops. Traditionally, agriculture, food science and technology, economics, nutrition, and sociology were separate disciplines. However, the newly designed system has combined all of these activities into one compressive larger unit termed a "food system". As a result, the undernourished population around the world is declining, from approximately 1 billion in 1991 to 792 million today [28]. Is it possible to use the same approach to prevent obesity?

Indeed, experts are proposing a holistic approach to prevent obesity. Proposed approaches include (1) increasing the diversity of locally available foods though food hub, (e.g., developing home gardens to be managed by women as a source of highly nutritious food for their families), (2) diet diversification with whole foods (pulse crops, fruits, and vegetables) and reduction of the intake of foods rich in refined sugar and fat, (3) developing appropriate technologies to preserve and store foods for local communities, (4) population-based guidelines to control food intake, e.g., food-based dietary guidelines (eat more fruits, vegetables, and legumes) or a nutrient-based dietary approach (10% energy from protein, 15–30% energy from fat, and >50% energy from complex carbohydrates), along with limited salt and alcohol intake, and (5) population-based guidelines for physical activity and a greater focus on basic social and biological research to understand household decision-making, food habits, child care, and food purchasing power [29, 31]. Notably, social level solutions for obesity prevention require attention from national government, food supply, media, non-government organizations, healthcare services, workforce, education, neighborhoods, homes, and families [29, 31].

3. The potential role of pulse crops in combating obesity

Diet modulates local and systemic inflammatory markers, suggesting a mechanism of action on obesity, diabetes, and other related conditions. In a fairly detailed study, a healthy diet, characterized by fruits, vegetables, tomato, poultry, pulses, tea, fruit juices, and whole grains, was inversely associated with inflammatory markers C-reactive protein, TNF-α, soluble vascular cell adhesion molecules (sVCAM-1), and E-selectin; conversely, a diet rich in refined grains, red meat, butter, processed meat, high-fat dairy, sweets, pizza, potato, and soft drinks was positively associated with systemic inflammation [32]. In particular, prebiotic food ingredients can decrease systemic inflammation through associations with hindgut microflora [33]. Pulse crops, and lentil in particular, are key whole foods that provide significant nutritional benefits in terms of micronutrient and prebiotic carbohydrate content and could thus play a role in obesity prevention.

3.1. Lentil production

Global lentil production has increased six-fold since the 1960s, from 0.85 million tons (Mt) in 1961 to 4.98 Mt in 2013; this has been accompanied by a 150% increase in sown area and a more than doubling of average yields from 528 to 1150 kg ha^{-1} (**Table 2**) [28]. Lentil is currently grown in as many as 51 countries, including Canada, India, Turkey, Australia, the United States, Nepal, China, Ethiopia, Syria, and Bangladesh. Although the area under lentil cultivation in Turkey has declined in the last decade, lentil growing areas in Australia, Canada, Ethiopia, India, Nepal, and the United States have considerably increased. Production in Asia is concentrated in a belt stretching from Turkey in the west to Bangladesh in the east, accounting for ~58% of the world sown area. China has recently started releasing lentil-related data, and Bangladesh has increased its productivity through release and cultivation of improved varieties. During the 1990s, Iran, Nepal, and Syria substantially increased production but lentil-cultivated area and production in Pakistan declined. Forty percent of the world's lentil production is in North America, where Canada and the United States are major producers and Mexico is a minor producer. In Africa, Ethiopia and Morocco are significant producers, while Algeria, Sudan, Egypt, and Tunisia are only minor producers. In South America, Argentina and Peru are the only significant producers. European lentil production is gradually decreasing, with France and Spain being the only noteworthy producers. In Oceania, Australia has emerged as a global leader with a production of 324,100 tons in 2013.

Year	Area (million ha)	Production (Mt)	Yield (kg/ha)
1961	1.62	0.85	528
1971	1.72	1.05	611
1981	2.27	1.45	640
1991	3.27	2.66	814
2001	3.99	3.25	816
2013	4.33	4.98	1,150

Reproduced from ref [28].

Table 2. Trend in world lentil production (1961–2013).

Exports account for approximately one-third of total lentil production, with the remainder eaten locally. International trade in small-seeded, red cotyledon lentil is dominated by Australia, Canada, and Turkey, whereas trade in large-seeded, green lentil is primarily led by Canada and the United States. Countries in the Indian subcontinent and the Middle East are major importers of red lentil, and southern Europe and South America import large-seeded green lentils. Lentils have been a popular food in many countries for thousands of years, likely because of the associated nutritional benefits and preference for vegetable protein over other protein sources.

3.2. Nutritional quality of lentils

Lentils contain a relatively high concentration of protein (~30%) and are a rich source of essential micronutrients (e.g., folates, iron, zinc, selenium, and carotenoids; **Table 3**). Lentils are naturally high in iron, zinc, and selenium in forms that are highly bioavailable to humans [42]. In addition, broad-sense heritability estimates for these minerals are high, indicating it is possible to breed lentil cultivars with enhanced ability to accumulate iron, zinc, and selenium in the seed despite environmental influences [37, 38]. Interestingly, lentils are very low in phytic acid, which enables greater mineral absorption in the human gut [39, 41]. Furthermore, bioavailability studies using both cultured Caco-2 cells and humans clearly show certain lentil varieties are rich in bioavailable iron and selenium [42, 43]. Mineral bioavailability in lentil is relatively high compared with cereals and other legumes because of the presence of high levels of iron absorption promoters (e.g., ascorbic acid) and low levels of Fe absorption inhibitors (e.g., phytic acid, gallic acid, and chlorogenic acid) [39, 42]. Lentil has been used as a model candidate crop for micronutrient biofortification research at the International Centre for Agricultural Research in the Dry Areas (ICARDA) in collaboration with the Clemson University Pulse Quality and Nutrition program. Our recent work has selected a few selenium-enriching lentil accessions to develop high selenium uptake lentil populations with the aim of releasing new lentil cultivars that produce seed with high bioavailable selenium [44]. Further studies are in progress with respect to the release of high iron and zinc lentil cultivars for India and Africa.

Nutrient component (units)	Concentration	Reference
Moisture (%)	1–12	[34]
Protein (%)	20–29	[34]
Ash (%)	1.8–3.3	[34]
Total lipid (fat) (%)	1–2	[35]
Carbohydrate, by difference (%)	60–63	[36]
Total starch (%)	40–70	[10]
Total prebiotic carbohydrates (g/100 g)	12.3–14.1	[10]
Resistant starch	5.5–9.3	[10]
Calcium, Ca (mg/kg)	460–496	[34]
Iron, Fe (mg/kg)	73–90	[37]
Potassium, K (mg/kg)	6,954–7,761	[34]
Zinc, Zn (mg/kg)	44–54	[37]
Selenium (μg/kg)	425–673	[38]
Ascorbic acid (mg/kg)	61.2–84.3	[39]
Gallic acid (mg/kg)	28.2–39.3	[39]
Chlorogenic acid (mg/kg)	10.3–20.3	[39]
Folate, dietary folate equivalent (μg/g)	2.2–2.9	[40]
Phytic acid (mg/g)	2.4–4.4	[41]
Fe bioavailability (ng/mg of protein)	7.2–22	[42]

Table 3. Nutritional composition of raw lentils.

4. Prebiotic carbohydrates in lentil

Prebiotics, also known as low digestible carbohydrates, are defined as *a selectively fermented ingredient that allows specific changes, both in the composition and/or activity in the gastrointestinal microflora that confers benefits upon host well-being and health* [45]. A prebiotic is a specific colonic nutrient that provides a metabolic substrate, and/or acts as a biosynthetic precursor or cofactor for human microbiota. Classification of a food as a prebiotic requires scientific demonstration that the ingredient (1) resists digestive processes in the upper part of the gastrointestinal tract, (2) is fermented by intestinal microbiota, and (3) selectively stimulates growth and/or activity of health promoting bacteria in that microbiotic population [45].

Food	Prebiotic type	Range (g/100 g)	Reference
Lentil	Sugar alcohol		
	Sorbitol	1.04–1.35	[10]
	Mannitol	0.16–0.29	[10]
	Galactinol	0.03–0.13	[10]
	Non-starch polysaccharides		
	RFO	2–4	[10]
	Nystose	0.06–0.07	[10]
	FOS	0.20–0.10	[46]
	Starch polysaccharides		
	Total starch	44–49	[10]
	Resistant starch	3.4–9.3	[10, 47]
Common bean	Non-starch polysaccharides		
	RFO	0–1.4	[46]
	FOS	0–0.5	[46]
	Starch polysaccharides		
	Resistant starch	1.8–1.9	[47]
White bread	Non-starch polysaccharides		
	RFO	0–0.2	[46]
	FOS	0–0.7	[46]
	Starch polysaccharides		
	Resistant starch	0.1–1.2	[47]
Jerusalem artichoke	Non-starch polysaccharides		
	FOS	12.2	[48]

Table 4. Types and concentrations of prebiotic carbohydrates in foods.

Lentils, on average, contain a total of 63% carbohydrates [36] and support healthful hindgut microflora [10]. Naturally occurring prebiotic carbohydrates in lentils are categorized into two major groups: (1) dietary fiber and (2) sugar alcohols [10, 45]. Dietary fiber is comprised of starch polysaccharides, including resistant starch (RS), and non-starch polysaccharides, including raffinose family oligosaccharides (RFO), fructooligosaccharides (FOS), galactooligosaccharide (GOS), xylooligosaccharide (XOS), and insulin. Sugar alcohols include sorbitol, mannitol, and galactinol. Types of prebiotic carbohydrates present in lentils and other foods are shown in **Table 4**. Concentration of prebiotics in foods range from trace levels, as seen in white bread, to relatively high amounts in Jerusalem artichoke. Many of these naturally occurring prebiotic carbohydrates are found in vegetables, pulses, and fruits at concentrations ranging from 35.7 to 47.6 g/100 g in chicory root to trace amounts in many vegetables [49]. In addition to vegetables and legumes, wheat, onion, and green bananas are other major sources of these carbohydrates. A few studies have reported RFO [50, 51] and RS [52] concentrations in lentil but did not assess genetic and environmental influences to determine the baseline concentration of these prebiotic carbohydrates in current lentil cultivars in production. Therefore, the ICARDA-Clemson University research program is working toward answering the following questions with respect to lentil prebiotic carbohydrates: *(1) What is the profile of lentil prebiotic carbohydrates?* [10] *(2) Is there any genetic or environmental variation in prebiotic carbohydrates in lentil?* [10, 53] *(3) What is the effect of dehulling, cooking, and cooling on lentil prebiotic carbohydrate concentrations?* [54] *and (4) What is the effect of lentil prebiotics on obesity?* The remainder of this chapter will provide an overview of research findings to date with respect to the first three questions as well as research currently in progress to address question four.

4.1. What is the profile of lentil prebiotic carbohydrates?

To understand the profile of prebiotic carbohydrates, we analyzed 10 commercial lentil genotypes grown in North Dakota, USA, in 2010 and 2011. Study results clearly characterized the following lentil prebiotics: FOS (kestose and nystose), RFO (raffinose, stachyose, and verbascose), sugar alcohols (sorbitol and mannitol), and total and resistant starch [10]. Mean concentrations of RFO, sugar alcohols, FOS, and resistant starch were 4071 mg, 1423 mg, 62 mg, and 7.5 g 100 g^{-1}, respectively. Total starch ranged from 45 to 48 g 100 g^{-1}. This means a 100 g serving of lentil could provide over 13 g of total prebiotic carbohydrates (**Table 3**) [10]. Overall, these results indicate that lentil contains nutritionally significant amounts of prebiotic carbohydrates, the levels of which could potentially be further increased through genetic selection and location sourcing.

4.2. Is there any genetic or environmental variation in prebiotic carbohydrates?

Work to date clearly demonstrates that prebiotic concentrations in lentils vary with genetic and environmental factors. Growing location/country significantly influences the concentration of various prebiotic carbohydrates (**Figure 2**). We completed a global survey of 335 lentil samples from 10 locations in 6 countries for sugar alcohols and various mono-, di-, and oligosaccharides, including RFO and FOS [54]. Mean LMWC concentrations varied widely: sorbitol, 1250–1824 mg/100 g; mannitol, 57–132 mg/100 g; galactinol, 46–89 mg/100 g; sucrose, 1,750–2,355 mg/100 g; raffinose + stachyose, 3,314–4,802 mg/100 g; verbascose, 1,907–2,453 mg/100 g; nystose, 8–450 mg/100 g; and kestose, from not detected to 244 mg/100 g. In addition, the concentrations of these prebiotics varied with average temperature and precipitation of the region/country of origin, which was expected because of the fact that sugar alcohols, RFOs, and sucrose are primarily stored in plants as reserves for survival during abiotic stress [55].

Moroccan lentils had consistently higher concentrations of RFOs and sugar alcohols (**Figure 2**) than those grown in other regions. Likewise, concentrations in Ethiopian and the American (WA) lentils were also high in RFOs. Regions with less precipitation and higher temperatures during the growing season showed higher concentrations of prebiotic carbohydrates, reflecting the existence of a response mechanism to water-deficit stress.

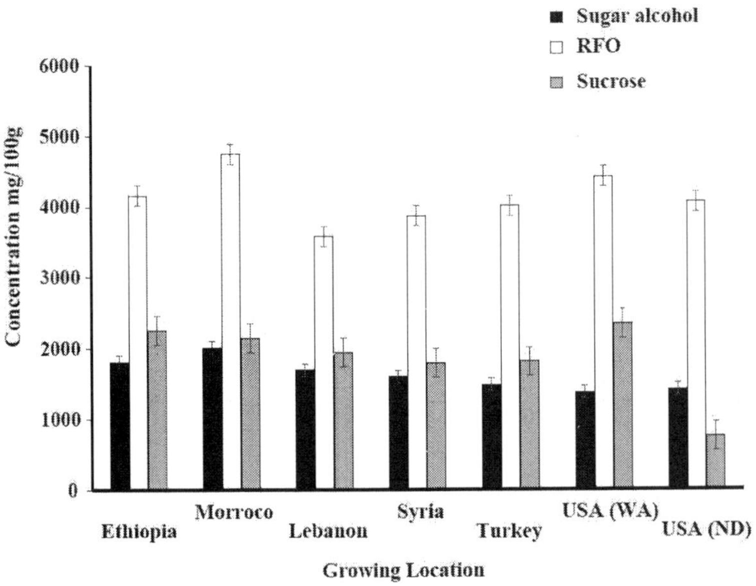

Figure 2. Sugar alcohol, raffinose family oligosaccharides (RFO), and sucrose concentration of lentils grown in seven different locations [10, 54].

Genotypic variation with respect to prebiotic carbohydrates was also evident. In lentils from the United StatesA [10], for example, total sugar alcohol concentrations were highest in CDC Riveland (1,598 mg 100 g^{-1}), RFO concentrations ranged from 3,508 mg 100 g^{-1} in CDC Rosetown to 4,652 mg 100 g^{-1} in Pennell, and FOS ranged from 52 mg 100 g^{-1} in CDC Red Rider to 79 mg 100 g^{-1} in CDC Viceroy. Resistant starch ranged from 6.0 to 8.9 g 100 g^{-1} and total starch ranged from 45 to 48 g 100 g^{-1}.

4.3. What is the effect of dehulling, cooking, and cooling on lentil prebiotic carbohydrate concentrations?

Food processing is an important component of food production. Generally, processing involves separation of non-edible parts, making foods more storage stable, and converting foods into easily cooked forms. Lentil processing involves three steps: cleaning, dehulling, and splitting. Cleaning removes other seeds, soils, and physical contaminants from the harvested lentils, dehulling involves removal of the lentil seed coats, and splitting breaks the lentil cotyledons into two halves. The last two steps in lentil processing are determined by consumer preference: some lentil consumers favor lentils with the seed coat intact while others choose non-split and split lentils without seed coats. Lentils can also be further processed to separate protein, carbohydrate, and other fractions, but this is not typical; more than 90% of lentil produced in the world is consumed in either whole seed with seed coat, football (dehulled but whole), or split form.

Lentil is a fast cooking food (~10 minutes), and the majority of consumers eat lentils as a soup or curry as a nutritional complement to accompanying rice or bread. Their short cooking time makes them a popular whole food for millions of people. Unlike other legumes, cooked lentils retain all of their nutritional value as there are no cooking water nutrient losses. As noted above, lentils can be processed into various fractions, with the carbohydrates and proteins contributing unique physical and chemical properties to foods when added as an ingredient. However, separation of lentil into different fractions requires higher energy and labor inputs that increase in cost of production. Such further lentil processing may also lead to a loss of essential nutrients found only in the whole seeds. Current lentil supply and demand, the increased focus on micronutrient delivery from whole foods, and consumer preferences suggest lentil will continue to be produced and marketed primarily as a whole food.

Few studies have indicated the effects of processing and cooking on lentil prebiotic carbohydrate concentrations. Two commercially available lentil market classes (medium green and small red) showed some RFO reductions with cooking, cooling, and reheating [54]. Mean RS concentrations in raw, cooked, cooled, and reheated lentil were measured at 3.0, 3.0, 5.1, and 5.1% (w/w), respectively, indicating cooling-induced synthesis of RS from gelatinized starch. These results highlight the impact of temperature on lentil nutritional quality and shows lentil are more nutritious after cooling. Similar increases (400%) in RS from raw to cooked then cooled potatoes have also been demonstrated [56].

4.4. What is the effect of lentil prebiotics on obesity?

Recent discoveries suggest the intestinal microbiome and a prebiotic-rich, low-caloric diet can play important roles in combating obesity and related diseases. Three dominant phyla have been identified in human fecal flora: Furmicutes, Bacteroides, and Actinobactteria. Sub-dominant groups are enterobacteria, streptococci, and lactobacilli [57]. The relative proportion of Bacteroidetes is decreased in obese individuals compared with lean individuals; however, this relative proportion rebounds with weight loss on a prebiotic-rich, low-caloric diet. Furthermore, consumption of non-digestible, fermentable carbohydrates (or prebiotics) may stimulate the growth and activity of hind gut bacteria by producing short-chain fatty acids that provide energy source for colonocytes, strengthen the gut mucosal barrier, and suppress colonization of pathogens [43]. As clearly shown by our research, lentil is a rich source of prebiotic carbohydrates, therefore, lentils offer new opportunities in this regard. To date, no research has been carried out to understand the true effect of lentil prebiotics on obesity. We are expecting to finish preliminary research related to this question in summer 2016. Lentil prebiotic carbohydrates provide numerous positive benefits to human health. However, accurate prebiotic carbohydrate characterization and quantification is important not only to determine types and levels but also inform consumers of food sources that are rich in these important compounds.

4.5. Accurate measurement of prebiotics

Human intestinal enzymes are able to digest carbohydrates based on their molecular structure. Carbohydrates can be categorized in to readily-, slow- and non-digestible carbohydrates, which are known chemically as mono-, di-, and polysaccharides. Glucose, a monosaccharide, is a readily digestible carbohydrate; however, starch is a glucose polysaccharide with a digestibility varying from fully digestible to non-digestible. Some prebiotic carbohydrates resist the activity of human digestive enzymes and pass to the large intestine where they are acted upon by bacterial enzymes. The use of both modern analytical instrumentation and enzymatic procedures is required to accurately characterize the true levels of prebiotic carbohydrates.

Carbohydrate analysis involves two main steps: carbohydrate isolation from a sample and analysis of those isolated compounds. Hot water extraction or a combination of hot water and ethanol is used to isolate most water-soluble carbohydrates. These water-ethanol soluble carbohydrates are small in molecular size. The most accurate analytical method to quantify these smaller carbohydrates is high performance anion exchange chromatography with pulsed amperometric detection (HPAE-PAD). Sugar alcohols of monosaccharides represent two groups of water-soluble carbohydrates that can be accurately identified and quantified by HPAE-PAD. Other carbohydrates such as disaccharides (two monosaccharide units) and oligosaccharides (3–10 monosaccharide units) are also water or water-ethanol mixture soluble carbohydrates and can be successfully be quantified using HPAE-PAD. However, HPAE-PAD and other analytical methods do not provide accurate identification and quantification of polysaccharides comprised of 10 or more monosaccharides. These carbohydrates can only be quantified only after hydrolyzing them into monosaccharides by enzymatic hydrolysis.

Enzyme hydrolysis of prebiotic carbohydrates has two objectives: (1) to simulate human intestinal environment to isolate digestible and non-digestible carbohydrate fractions and (2) to selectively hydrolyze the glycosidic linkage of carbohydrates for accurate quantification. For example, amylases are a type of human digestive enzyme that can be used to separate starches from resistant starches during sample preparation/isolation procedures. The same type of enzyme can also be used to completely hydrolyze starch-like macromolecules into their simpler glucose units, thus enabling quantification of larger carbohydrate molecules.

Sugar alcohols such as sorbitol have the simplest molecular structure of all prebiotic carbohydrates; however, resistant starches are large, complex macromolecules. Well-established HPAE-PAD and enzymatic procedures are available to accurately quantify some but not all prebiotic carbohydrates. However, work is ongoing to combine good understanding of prebiotic carbohydrate molecular structures, HPAE-PAD instrumentation, and enzymatic hydrolysis procedures to accurately quantify more prebiotic carbohydrates.

5. Lentil breeding at the ICARDA

Within the biofortification framework, the ICARDA lentil breeding program is working together with Clemson University to increase mineral concentration and bioavailability to combat global micronutrient malnutrition. In addition, future lentil selections will be carried out by selecting cultivars with higher micronutrients, prebiotic carbohydrates, and low phytate. Biofortification can improve crop nutritional value with minimal impact on consumer cost. The ICARDA has also created a composite collection of more than 1,000 lentil lines to understand the genetic diversity with respect to different nutritional traits, including iron and zinc accumulation. Knowledge of the patterns of variation in the world germplasm collection for yield, disease resistance, and nutrition is the key to understand factors affecting lentil adaptation that can then be applied to lentil breeding. The geographic distribution of these landraces in the world lentil collection for various morphological characteristics, responses in flowering to temperature and photoperiod, winter hardiness, iron-deficiency chlorosis, and boron imbalances collectively illustrate the specificity of adaptation in lentil. Additional information on the specificity of adaptation within the crop has come from collaborative multi-environment yield trials of common entries selected in different locations.

Understanding genotypes and environmental factors, local constraints to production, and the various consumer requirements of different geographic areas has led the breeding program at the ICARDA to develop new genetic materials for a series of separate but finely targeted geographical streams, linked closely to national breeding programs. The major agro-ecological regions of production of lentil being targeted include (1) South Asia, East Africa, and Yemen, (2) low-to-medium elevation Mediterranean regions, and (3) high-elevation areas of West Asia and North Africa. These regions correspond to early, medium, and late maturity groups, respectively. Additionally, lentil improvement activities have also been extended to the Central Asia and Caucasus (CAC) region, where an initial thrust has been to study the adaptation of diverse material suitable to their agro-climatic conditions.

6. Final thoughts

Recent studies have focused on health-beneficial bioactive components in commonly eaten foods to understand their impact on human health and disease. Among these bioactive compounds, prebiotic carbohydrates are important food constituents to reduce obesity-related non-communicable diseases through interactions with the hindgut microbiome. Lentils induce a low-glycemic response, and this has been attributed to their prebiotic content that has a high resistance to human enzyme hydrolysis. In addition to human health benefits, prebiotic carbohydrates are important for plant survival, e.g., with respect to water-deficit or cold stress. Overall, lentil is a highly nutritious pulse crop that has supplied essential nutrients including proteins, dietary fiber, macro, and micronutrients to various populations over centuries.

Despite current research evidence on prebiotic health effects, lentil breeding programs continue to work toward lentil cultivars with reduced levels of RFOs in response to consumer preference in certain markets [51]. RFOs have long been known as antinutrients and linked to the flatulence and gastrointestinal discomfort that occur in some consumers [58]. Conventional plant breeding programs have reduced the RFO concentration in seeds [59]; however, human health benefits associated with RFOs have begun to be documented. RFO and other prebiotic carbohydrates are important dietary components for preventing overweight and obesity and associated diseases. Therefore, the aim of future plant-breeding efforts may instead be to enhance all types of prebiotic carbohydrates.

Acknowledgements

Funding support for the Pulse Quality and Nutrition research program is provided by Clemson University, SC, USA; the International Centre for Agricultural Research in the Dry Areas (ICARDA), Rabat, Morocco; the USA Dry Pea Lentil Council; and the American Pulse Association, WA, USA.

Author details

Dil Thavarajah[1*], Pushparajah Thavarajah[1,2], Casey R. Johnson[3] and Shiv Kumar[4]

*Address all correspondence to: dthavar@clemson.edu

1 Agricultural and Environmental Sciences, Clemson University, Clemson, SC, USA

2 International College Beijing, China Agricultural University, Beijing, China

3 Mayo Medical School, Rochester, MN, USA

4 Biodiversity and Integrated Gene Management Program, International Centre for Agricultural Research in the Dry Areas (ICARDA), Rabat-Institute, Rabat, Morocco

References

[1] United Nations, 2012. Prevention and control of non-communicable diseases. Report of the Secretary General. http://www.un.org/en/ga/president/65/issues/ncdiseases.shtml. Accessed: 2015-12-5.

[2] World Health Organization, 2015. Controlling the global obesity epidemic. http://www.who.int/nutrition/topics/obesity/en/. Accessed: 2015-12–5.

[3] Ogden CL, Carroll MD, Kit BK, Flegal KM. Prevalence of obesity in the United States, 2009–2010. NCHS Data Brief. 2012, No. 82.

[4] Caselatod SVM, Freitasd S, Sgarbieri VC. The importance of prebiotics in functional foods and clinical practice. Food and Nutrition Sciences. 2011; 2 (2), 133–144.

[5] Manning TS, Gibson GR. Prebiotics. Best Practice and Research in Clinical Gastroenterology. 2004; 18 (2), 287–298.

[6] Lee YK, Mazmanian SK. Has the microbiota played a critical role in the evolution of the adaptive immune system? Science. 2010; 330 (6012), 1768–1773.

[7] Burns AJ, Rowland IR. Anti-carcinogenicity of probiotics and prebiotics. Current Issues in Intestinal Microbiology. 2000;1 (1), 13–24.

[8] Conlon MA, Kerr CA, Mc Sweeney CS, Dunne RA, Shaw JM, Kang S, Bird AR, Morell MK, Lockett TJ, Molloy PL, Regina A, Toden S, Clarke JM, Topping DL. Resistant starches protect against colonic DNA damage and alter microbiota and gene expression in rats fed a western diet. Journal of Nutrition. 2012; 142 (5), 832–840.

[9] Johnston KL, Thomas EL, Bell JD, Frost GS, Robertson MD. Resistant starch improves insulin sensitivity in metabolic syndrome. Diabetic Medicine. 2010; 27 (4), 391–397.

[10] Johnson CR, Thavarajah D, Combs Jr. GF, Thavarajah P. Lentil (*Lens culinaris* L.): A prebiotic-rich whole food legume. Food Research International. 2013; 51(1), 107–113.

[11] Swinburn BA, Sacks G, Hall KD, Mc Pherson K, Finegood DT, Moodie ML, et al. The global obesity pandemic: Shaped by global drivers and local environments. Lancet. 2011; 378 (9793), 804–814.

[12] Popkin BM, Adair LS, Ng SW. Global nutrition transition and the pandemic of obesity in developing countries. Nutrition Reviews. 2012; 70 (1), 3–21.

[13] Ng M, Fleming T, Robinson M, Thomson B, Graetz N, Margono C, et al. Global, regional, and national prevalence of overweight and obesity in children and adults during 1980–2013: A systematic analysis for the Global Burden of Disease Study 2013. Lancet. 2014; 384(9945), 766–781.

[14] Lim SS, Vos T, Flaxman AD, Danaei G, Shibuya K, Adair-Rohani H, et al. A comparative risk assessment of burden of disease and injury attributable to 67 risk factors and risk factor clusters in 21 regions, 1990–2010: A systematic analysis for the Global Burden of Disease Study 2010. Lancet. 2012; 380(9859), 2224–2260.

[15] Burton BT, Foster WR, Hirsch J, Van Itallie TB. Health implications of obesity: An NIH Consensus Development Conference. International Journal of Obesity. 1985; 9 (3), 155–170.

[16] Must A, Spadano J, Coakley EH, Field AE, Colditz G, Dietz WH. The disease burden associated with overweight and obesity. JAMA. 1999; 282 (16), 1523–1529.

[17] Withrow D, Alter DA. The economic burden of obesity worldwide: A systematic review of the direct costs of obesity. Obesity Reviews. 2011; 12 (2), 131–141.

[18] Gortmaker SL, Must A, Perrin JM, Sobol AM, Dietz WH. Social and economic consequences of overweight in adolescence and young adulthood. New England Journal of Medicine. 1993; 329 (14), 1008–1012.

[19] Katz DA, Mc Horney CA, Atkinson RL. Impact of obesity on health-related quality of life in patients with chronic illness. Journal of General Internal Medicine. 2000; 15 (11), 789–796.

[20] Health UNIo. Clinical guidelines on the identification, evaluation, and treatment of overweight and obesity in adults: Executive summary. Expert Panel on the Identification, Evaluation, and Treatment of Overweight in Adults. American Journal of Clinical Nutrition. 1998; 68 (4), 899–917.

[21] Weinsier RL, Hunter GR, Heini AF, Goran MI, Sell SM. The etiology of obesity: Relative contribution of metabolic factors, diet, and physical activity. American Journal of Medicine. 1998; 105 (2), 145–150.

[22] Mc Carthy MI. Genomics, Type 2 diabetes, and obesity. New England Journal of Medicine. 2010; 363 (24), 2339–2350.

[23] Bastard J-P, Maachi M, Lagathu C, Kim MJ, Caron M, Vidal H, et al. Recent advances in the relationship between obesity, inflammation, and insulin resistance. European Cytokine Network. 2006; 17 (1), 4–12.

[24] Linton MF, Fazio S. Macrophages, inflammation, and atherosclerosis. International Journal of Obesity. 2003; 27, S35–S40.

[25] Mathieu P, Lemieux I, Després JP. Obesity, inflammation, and cardiovascular risk. Clinical Pharmacology & Therapeutics. 2010; 87 (4), 407–416.

[26] Rocha VZ, Libby P. Obesity, inflammation, and atherosclerosis. Nature Reviews Cardiology. 2009; 6 (6), 399–409.

[27] Cai D, Yuan M, Frantz DF, Melendez PA, Hansen L, Lee J, et al. Local and systemic insulin resistance resulting from hepatic activation of IKK-β and NF-κB. Nature Medicine. 2005; 11 (2), 183–190.

[28] FAOSTAT, 2015. Food and agricultural commodities production. http://faostat.fao.org/site/. Accessed: 2015-12-5.

[29] Kumanyika S, Jeffery RW, Morabia A, Ritenbaugh C, Antipatis VJ. Public health approaches to the prevention of obesity (PHAPO) working group of the international obesity task force (IOTF). International Journal of Obesity. 2002; 26, 425–436.

[30] Welch RM, Graham RD. Agriculture: The real nexus for enhancing bioavailable micronutrients in food crops. Journal of Trace Elements in Medicine and Biology. 2005; 18 (4), 299–307.

[31] Sacks G, Swinburn BA, Lawrence MA. A systematic policy approach to changing the food system and physical activity environments to prevent obesity. Australia and New Zealand Health Policy. 2008; 5 (1), 13.

[32] Esmaillzadeh A, Kimiagar M, Mehrabi Y, Azadbakht L, Hu FB, Willett WC. Dietary patterns and markers of systemic inflammation among iranian women. Journal of Nutrition. 2007; 137 (4), 992–998.

[33] Delzenne NM, Neyrinck AM, Bäckhed F, Cani PD. Targeting gut microbiota in obesity: Effects of prebiotics and probiotics. Nature Reviews Endocrinology. 2011; 7 (11), 639–646.

[34] Pulse Quality Survey. 2013. http://www.pea-lentil.com/core/files/pealentil/uploads/files/2013%20U_S_%20Pulse%20Quality%20Survey.pdf. Accessed: 12-5-2015.

[35] USDA. 2015. USDA Nutrient data base for standard reference. http://ndb.nal.usda.gov/. Accessed: 12-5-2015.

[36] Bhatty RS. Composition and quality of lentil (*Lens culinaris* Medik.)—A review. Canadian Institute of Food Science and Technology Journal. 1988; 21, 144–160.

[37] Thavarajah D, Thavarajah P, Sarker A, Vandenberg A. Lentils (*Lens culinaris* Medikus Subspecies culinaris): A whole food for increased iron and zinc intake. Journal of Agricultural and Food Chemistry. 2009; 57 (12), 5413–5419.

[38] Thavarajah D, Ruszkowski J, Vandenberg A. High potential for selenium biofortification of lentils (*Lens culinaris* L.). Journal of Agricultural and Food Chemistry. 2008; 56 (22), 10747–10753.

[39] Johnson CR, Thavarajah D, Thavarajah P. The influence of phenolic and phytic acid food matrix factors on iron bioavailability potential in 10 commercial lentil genotypes (*Lens culinaris* L.). Journal of Food Composition and Analysis. 2013; 31 (1), 82–86.

[40] Sen Gupta D, Thavarajah D, Knutson P, Thavarajah P, Mc Gee RJ, Coyne CJ, Kumar S. Lentils (*Lens culinaris* L.), a rich source of folates. Journal of Agricultural and Food Chemistry. 2013; 61 (32), 7794–7799.

[41] Thavarajah P, Thavarajah D, Vandenberg A. Low phytic acid lentils (*Lens culinaris* L.): A potential solution for increased micronutrient bioavailability. Journal of Agricultural and Food Chemistry. 2009; 57 (19), 9044–9049.

[42] Thavarajah D, Thavarajah P, Wejesuriya A, Rutzke M, Glahn RP, Combs GFJr, Vandenberg A. The potential of lentil (*Lens culinaris* L.) as a whole food for increased selenium,

iron, and zinc intake: Preliminary results from a 3 year study. Euphytica. 2011; 180 (1), 123–128.

[43] Migliozzi M, Thavarajah D, Thavarajah P, Smith P. Lentil and kale: Complementary nutrient-rich whole food sources to combat micronutrient and calorie malnutrition. Nutrients. 2015; 7 (11), 9285–9298.

[44] Thavarajah D, Thavarajah P, Throughl E, Gebhardt M, Lacher C, Kumar S, Combs GF. Will selenium increase lentil (*Lens culinaris* Medik) yield and seed quality? Frontiers in Plant Science. 2015; 6, 356.

[45] Roberfroid MB. Prebiotics: The concept revisited. Journal of Nutrition. 2007; 137, 830S–837S.

[46] Biesiekierski JR, Rosella O, Rose R, Liels K, Barrett JS, Shepherd SJ, et al. Quantification of fructans, galacto-oligosacharides and other short-chain carbohydrates in processed grains and cereals. Journal of Human Nutrition and Dietetics. 2011; 24, 154–176.

[47] Murphy MM, Douglass JS, Birkett A. Resistant starch intakes in the United States. Journal of the American Dietetic Association. 2008; 108, 67–78.

[48] Muir JG, Shepherd SJ, Rosella O, Rose R, Barrett JS, Gibson PR. Fructan and free fructose content of common Australian vegetables and fruit. Journal of Agricultural and Food Chemistry. 2007; 55, 6619–6627.

[49] van Loo J, Coussement P, de Leenheer L, Hoebregs H, Smits G. On the presence of inulin and oligofructose as natural ingredients in the Western diet. Critical Reviews in Food Science and Nutrition, 1999; 35 (6), 525–552.

[50] Wang N, Hatcher DW, Toews R, Gawalko EJ. Influence of cooking and dehulling on nutritional composition of several varieties of lentils (*Lens culinaris*). LWT—Food Science and Technology. 2009; 42, 842–848.

[51] Tahir M, Vandenberg A, Chibbar RN. Influence of environment on seed soluble carbohydrates in selected lentil cultivars. Journal of Food Composition and Analysis. 2011; 24, 596–602.

[52] Chung HJ, Liu Q, Dormer E, Hoover R, Warkentin TD, Vandenberg B. Composition, molecular structure, properties, and in vitro digestibility of starches from newly released Canadian pulse cultivars. Cereal Chemistry. 2008; 85, 473–481.

[53] Johnson CR, Thavarajah D, Thavarajah P, Fenlason A, Mc Gee R, Kumar S, Combs GF. A global survey of low-molecular weight carbohydrates in lentils. Journal of Food Composition and Analysis. 2015; 44, 178–185.

[54] Johnson CR, Thavarajah D, Thavarajah P, Payne S, Moore J, Ohm JB. Processing, cooking, and cooling affect prebiotic concentrations in lentil (*Lens culinaris* Medikus). Journal of Food Composition and Analysis. 2015; 38, 106–111.

[55] Eveland AL, Jackson DP. Sugars, signaling, and plant development. Journal of Experimental Botany. 2012; 63 (9), 3367–3377.

[56] Mishra S, Monro J, Hedderley D. Effect of processing on slowly digestible starch and resistant starch in potato. Starch – Stärke. 2008; 60, 500–507.

[57] Bajzer M, Seeley RJ. Physiology: Obesity and gut flora. Nature. 2006; 444 (7122), 1009–1010.

[58] Fleming SE. A study of relationships between flatus potential and carbohydrate distribution in legume seeds. Journal of Food Science. 1981; 46 (3), 794–798.

[59] Frias J, Bakhsh A, Jones DA, Arthur AE, Vidal-Valverde C, Rhodes MJ, Hedley CL. Genetic analysis of the raffinose oligosaccharide pathway in lentil seeds. Journal of Experimental Botany. 1999; 50 (333), 469–476.

Chapter 3

Pulse Proteins: From Processing to Structure-Function Relationships

Ashish Singhal, Asli Can Karaca, Robert Tyler and Michael Nickerson

Additional information is available at the end of the chapter

http://dx.doi.org/10.5772/64020

Abstract

Interest in alternative protein sources to those derived from animal, soy and wheat is on the rise, as consumers are searching for lower cost, healthier alternatives without compromising product quality and safety. Pulses are rich in protein, carbohydrates, vitamins and minerals and are low in fat. Although pea proteins experience greater integration into the plant protein ingredient market than others, lentil, chickpea, bean and faba beans are not far behind. This review discusses approaches used for extracting pulse proteins used to produce protein products (concentrates/isolates), mechanism driving structure-function relationships as well as potential applications.

Keywords: Pulse proteins, extraction, structure-function and applications, Legumin: Vicilin

1. Introduction

Pulses such as beans, peas and lentils have been consumed for thousands of years and represent one of the most extensively consumed food in the world [1]. Pulses play crucial roles in fulfilling the nutritional requirements of the growing population in a cost effective manner, especially for developing or underdeveloped countries where animal protein consumption is either limited or expensive [2]. Pulses are widely used for food purposes because of their high protein content, high nutritional and health beneficial properties, appropriate functional attributes, and associated low production cost and abundance [3]. The health benefits associated with pulse consumption include lowering of cholesterol levels, reducing the risks of various cardiovascular diseases and cancers, and decreasing the risk of type-2 diabetes [4]. Along with protein, pulses provides dietary fiber and vitamins and minerals such as iron, zinc, folate, and magnesium [1]. Pulses also have an antioxidant and anti-carcinogenic effect because of the presence of phytochemicals, saponins and tannins in them [1].

For many years, pulses have been used in the preparation of wholesome nutritional meals in combination with other food sources or ingredients. Pulse crops such as pea, chickpea and common bean (*Phaseolus vulgaris* L.), when blended with regionally grown cereal grains, could be of immense value in helping to fulfill the nutritional requirements of people relying just on mono-carbohydrate diets [5]. However, the nutritional quality of pulses is limited because of the presence of heat labile and heat stable anti-nutritional factors (ANFs) [2]. The ANFs include proteins such as lectins and protease inhibitors, and other compounds such as phytate, tannins, saponins, and alkaloids [2]. The negative impact of these ANFs on consumption of pulses in human and animal diets has been extensively reported [6]. However, the processed forms of legumes (flours, concentrates or isolates) are reported to have lower levels of ANFs than their corresponding raw material (seeds) [7]. For instance, during the germination process, legumes were found to have a higher digestibility, soluble protein [8] and dietary fiber [9, 10], and reduced levels of ANFs [11]. Furthermore, protein isolates prepared by extraction or precipitation methods were also found to have reduced anti-nutritional factors such as trypsin inhibitors, glycosides (such as convicine and vicine) and hemagglutinins which would otherwise impair protein digestion and could be toxic for human consumption [5, 12–14]. The exploitation of protein isolates or concentrates in new food formulations is of great importance because of their high nutrition and functionality [15]. The utilization of right individual functional properties might be useful in producing different food products such as cakes, biscuits, beverages and breads.

2. Protein structure and legumin/vicilin (L/V) ratio

The majority of pulse proteins are albumin and globulin fractions, where globulins represent ~70% and albumins constitute 10–20% of the total pulse protein [5, 16]. In addition, other proteins are present in minor proportions such as prolamins and glutelins [17, 18]. These four proteins can be classified according to their solubility in various solvents based on the Osborne classification scheme [19]. For example, globulin proteins are soluble in dilute salt solution, albumins in water, prolamins in 70% ethanol solution, and glutelins are solubilized in dilute alkali solutions [19, 20].

Albumins encompass structural and enzymatic proteins, lectins and protease inhibitors, with their overall molecular mass (MM) ranging between 5 and 80 kDa [5]. In contrast, the salt soluble globulins include legumin (11S, S = Svedberg Unit) and vicilin (7S) proteins. The 11S fraction is a hexamer (MM of ~340–360 kDa) comprised of six subunits (MM of ~60 kDa) linked by non-covalent interactions. Each subunit pair is comprised of an acidic (MM of ~40 kDa) and basic (MM of ~20 kDa) chain joined by a disulfide bond [16, 21]. In contrast, the 7S fraction is a trimer with a MM of ~175–180 kDa, and lacks disulfide bridging [5]. Vicilin protein

http://dx.doi.org/10.5772/64020

molecules also have been reported to have various subunits of 75, 43, 33, 56, 12 and 25 kDa [16, 21]. A third type of globulin is also present, although in lesser amounts as compared to other globulins, and is known as convicilin [22]. It is a 7S globulin, and a single convicilin molecule has an overall MM of 220–290 kDa, and consists of 3 or 4 subunits each with a MW of 70 kDa. This protein has a different amino acid profile than vicilin as it contains sulfur-containing amino acids, is immunologically similar to 7S vicilin, and contains very little carbohydrate [5]. Various pulse species have been reported to contain convicilin-type proteins. For example, Saenz de Miera et al. [23] investigated 29 different legume species from 4 genera (*Pisum, Lens, Vicia* and *Lathyrus spp.*), and reported the presence of 34 new convicilin gene sequences. All of the above studies considered convicilin as a third class of globulin molecules. However, O'Kane et al. [24] deny the consideration of convicilin as a third pea globulin based on their findings and reported that convicilin (a polypeptide) should be denoted as the R-subunit of pea vicilin molecules (salt extracted).

The ratio of legumin:vicilin (L/V) is not fixed and may vary among different pulse varieties and species. The ratio of L/V for pea, soybean and faba bean varies in the range of 0.2–8.0, 1.3–3.4 and 1.7–3.7, respectively [25–35]. Various studies reported that L/V ratio for wrinkled pea seeds (0.2–0.6) represents a smaller ratio compared to the smooth pea seeds (0.3–2.0) [28, 30, 35, 36]. Various factors including the methods used in the preparation of protein materials (concentrates or isolates), processing parameters like pH and temperature and environmental or agronomic factors may account for the variation in these ratios, which in turn could also have influential effects on the physiochemical properties of pulse protein materials [16, 21, 37, 38]. As a part of their studies, Barac et al. [38] extracted the proteins from six varieties (genotypes) of pea (Calvedon, L1, L2, L3, Maja and M.A) and indicated that genotypes with high 7S protein levels or low 11S protein levels yielded higher amounts of protein (protein extractability) compared to the other genotypes. Moreover, pure vicilin solutions were observed to have better functional properties (such as emulsification and gelation) than the pure legumin solutions [38]. It was indicated that a low L/V ratio for preparation of protein isolates could be desirable. In the Mertens et al. [35] study on smooth pea seeds, it was reported that agronomic factors, including variety, cultivar type and location, affected the protein content and L/V ratio with high significance. However, some varieties were less sensitive to the prevailing climatic conditions than others. This approach could be beneficial from an industrial point of view as it could manifest in picking stable and less sensitive L/V ratio lines for specific product quality characteristics [35].

Various groups have researched relationships between L/V ratios and their functional attributes. A number of studies noted that pea vicilin showed higher emulsifying properties than corresponding pea legumin [39–41], which was attributed due to higher solubility [42] and surface hydrophobicity [5] of vicilin proteins. Furthermore, Shen and Tang [43] reported that emulsifying properties of vicilins were found to be dependent on both the legume source (Kidney bean, red bean and mung bean) and their protein concentration (0.25–2.5% w/v). The differences in the emulsion properties of vicilins at different concentrations were majorly related to the variation in zeta potential and interfacial characteristics, and were also found to be dependent on other factors such as protein folding, penetration and structural rearrangement at the interface [43]. Bora et al. [44] studied the heat induced gelation of mixed pea globulins and found that 7S globulin had the capacity to undergo heat gelation while 11S globulin did not although used the same optimal conditions of gelation with 15% globulin solutions, pH 7.1 and heating at 87°C for 20 min. However, Nakamura et al. [45] observed that the gels formed by 7S globulins of soybean are less strong and transparent as compared to those formed by 11S globulins, which were much harder and turbid in nature. The study suggested that the extent of interaction in gel formation of a mixed system of 7S and 11S globulins is affected by factors such as the 11S/7S ratio and the composition of their subunits. Cserhalmi et al. [39] reported that mixed globulins and 7S fractions of pea proteins had increased surface hydrophobicity and emulsifying properties compared to the albumins and 11S fractions. Moreover, for all the pea varieties tested, the emulsifying and surface hydrophobicity properties were different from each other. Thus, varying the L/V ratio could be used in obtaining the desired functional attribute in new food formulations.

The quantification of 7S and 11S fractions present in isolates or concentrates is an essential step for calculation of L/V ratio which can be achieved using various methods described in literature. Methods include ammonium sulfate salt extraction [46], isoelectric precipitation [47], sodium dodecyl sulfate-polyacrylamide-gel electrophoresis (SDS-PAGE), gel chromatography [48], selective thermal denaturation [49], sucrose gradient centrifugation [50] and zonal isoelectric precipitation [51, 52]. The effective separation and the choice of technique should be dependent on factors such as nature of sample (isolates, concentrates, seed), extraction technique employed and the level of purification required. For testing of functional and physicochemical properties of 7S and 11S fractions, it is required that enough quantity of these samples is obtained whichever technique is used without compromising the purity.

3. Protein extraction

Protein extraction is dependent on many factors such as pH, temperature, particle size, ionic strength, type of salt used, and solvent to flour ratio [53, 54]. Various extraction methods are being studied so as to maximize the protein yield without compromising the protein functionality of the concentrate or isolate product. The protein extraction processes which are being exploited in the preparation of protein-rich materials (such as isolates and concentrates) can be classified into dry and wet methods [55–57].

3.1. Dry processing

Dry processing of pulses is typically done by air classification, which involves the separation of flours on the basis of particle size and density using an air stream into protein and starch rich fractions [21, 58]. Air classification has been found to be suitable for legume crops low in fat, such as field pea and common bean. Flours are first fractionated into starch (SI) and protein (PI) rich concentrates using an air classification method. SI is then remilled and fractionated to give SII and PII concentrates [55]. Protein separation efficiency (PSE) is defined as the percentage of total flour protein recovered in the PI and PII fractions, and measured as the subtraction of % total flour protein recovered in SII fraction from 100% [55]. For legume crops high in fat such as soybean and chickpea, particle agglomeration is detected which interferes with PSE [59–61]. Dry processing has major advantage over wet extraction methods as the native functionality of proteins is retained and a lower amount of energy and no water is required [62]. Moreover, in contrast to wet extraction methods where both protein concentrates and isolates can be produced, dry processes are suitable only for preparing protein concentrates with protein content from 40–75% [63] probably because of the presence of higher amount of other compounds such as oil and fibers, and protein loss in coarse fractions [64].

Tyler et al. [55] studied the fractionation of eight legumes (cowpea, great northern bean, lima bean, mung bean, navy bean, lentil, faba bean and field pea) using flours produced by pin milling followed by air classification and found faba bean (63.8–75.1%) and lima bean (43.4–49.6%) to have the highest and lowest protein concentrations in the protein-rich fractions. According to the authors, the suitability of pin milling followed by air classification is strongly correlated with the PSE of the legumes. Mung bean, lentil and great northern bean were found to have the highest mean PSE values of 88.9, 87.2 and 87.0%, respectively, whereas lima bean, cowpea and navy bean showed the lowest at 80.2, 78.2 and 80.3%, respectively. The other two legumes, faba bean and field pea, had PSE values of 84.1 and 82.8%, respectively. Overall, the authors indicated that except for lima bean and cowpea, the legumes were found to be suitable for separation of protein and starch fractions by the pin milling and air classification method.

3.2. Wet processing

In general, wet extraction methods can be exploited for preparing both protein concentrates and isolates at levels of 70% and 90% protein (or higher), respectively. However, it should be noted that currently there is no universal classification scheme which separates concentrate from an isolate for all the legumes. The various wet extraction processes include acid/alkaline extraction-isoelectric precipitation, ultrafiltration and salt extraction. Legume flours dispersed in aqueous solutions typically show high solubility when subjected to alkaline or acidic extraction conditions at pH 8–10 and below 4 respectively [63].

3.2.1. Acid/alkaline extraction-isoelectric precipitation (IEP)

Briefly, proteins are first dissolved under alkaline (alkaline extraction) or acidic (acid extraction) conditions, followed by a clarification step and then precipitation by adjusting the pH to the isoelectric point (pI) of the protein [65]. In solutions with the pH < pI, proteins assume a net positive charge, whereas at pHs > pI proteins assume a net negative charge. Under solvent conditions where proteins carry a net positive or negative charge, repulsive forces between proteins repel neighboring molecules, and also promote protein-water interactions for improved dispersion and solubility. Near the pI value, proteins tend to carry a neutral net charge, allowing neighboring proteins to aggregate via attractive van der Waals forces and hydrophobic interactions. Under these conditions, protein-protein interactions are favored over protein-water interactions, and thus protein is precipitated out of the solution.

According to Han and Hamaker [65], alkaline extraction followed by isoelectric precipitation is the most widely used method for obtaining extracts with protein purity greater than 70%. During alkaline extraction, legume proteins become solubilized at high pH values. The solution can then be clarified by centrifugation to remove insoluble material such as insoluble fiber, carbohydrates and insoluble proteins (e.g., prolamins). Protein concentrates or isolates can be formed by reducing the pH of the supernatant to near the pI of the protein using an acid such as HCl [63, 66]. The study of Can Karaca et al. [16] showed that isolates prepared from legumes (faba bean, chickpea, lentil, pea and soybean) by an alkaline extraction/IEP method had higher overall protein content (85.6%) as compared to those prepared by a salt extraction method (78.4%). Moreover, it was reported that both legume source and protein extraction method along with their interaction had significant effects on protein levels of the isolates, and also on physicochemical and emulsifying properties. The overall surface charge, solubility, hydrophobicity and creaming stability for IEP produced isolates was higher as compared to isolates produced by salt extraction [16]. The effect of processing or extraction conditions on the protein content of isolates can also be well observed from the studies of Flink and Christiansen [67] and McCurdy and Knipfel [68]. In the former study, faba bean isolates with protein contents of 80.0–90.0% were obtained when the bean:solvent ratio was 1:5 (w/v) with pH 8 to 10 at 23°C for 10 min, and the precipitation of protein was carried out at pH 3–5. While in the latter study, the protein content of faba bean isolates was 76.4–94.0% using a bean:solvent ratio of 1:5 w/v with pH 7–10, for 30 min, temperatures of 10°C and 20°C, and precipitation at pH 4–5.3.

Acid extraction (in principle similar to alkaline extraction) involves the preliminary extraction of proteins under acidic conditions. This process could result in high solubilization of proteins prior to protein recovery (IEP, Ultrafiltration (UF)), as proteins tend to be more soluble under acidic conditions (pH below 4) [5]. In a study by Vose [69] for preparation of faba bean (*Vicia faba equina* L. cv. Diana) and pea (*Pisum sativum* L. cv. Trapper) IEP isolates, the cyclone discharge obtained from pin milling these two legumes was acidified directly using 2 N HCl to a isoelectric point of 4.4–4.6. This process resulted in pea and faba bean protein isolates with 91.9% and 91.2% protein, respectively [5].

3.2.2. *Ultrafiltration/diafiltration*

In the literature, membrane separation methods were shown to produce protein isolates with higher functionality [70, 71] and were effective in reducing levels of anti-nutritional components which include protease and amylase inhibitors, lectins and polyphenols [72–74]. UF and microfiltration are membrane-based fractionation methods using pressure as the driving force for separation. Microfiltration can be used to separate particles or macromolecules larger than 0.1 μm, whereas ultrafiltration removes similar particles in the range of 0.001–0.02 μm [75]. For preparation of protein materials using ultrafiltration, the supernatant after alkaline or acidic extraction is processed using either UF or diafiltration (DF) together to isolate the protein material. UF is often combined with DF to improve protein recovery, where water is added to the retentate for dilution purposes, followed by re-ultrafiltration.

Vose [69] used the UF procedure to produce faba bean and pea protein isolates which protein levels of 94.1% and 89.5%, respectively. Boye et al. [66] evaluated the protein content of isolates obtained from different pulses (pea, chickpea and lentil) using alkaline extraction-IEP and UF/DF extraction methods. The protein content in concentrates obtained by the UF/DF method was found to be higher than in those obtained by IEP. For instance, for yellow pea, green lentil, red lentil, and desi and kabuli chickpea, UF/DF gave protein levels of 83.9%, 88.6%, 82.7%, 76.5% and 68.5%, respectively. In contrast, for IEP extraction, protein levels were 81.7%, 79.1%, 78.2%, 73.6% and 63.9% respectively for the same legume crops. Moreover, it was reported that UF was different from IEP in terms of protein composition as the isolates prepared by UF comprised both globulins and albumins, whereas the isolates prepared by IEP were observed to contain only globulins [63, 76, 77].

3.2.3. Salt extraction

Salt extraction is a process where globulin proteins are separated from albumins on the basis of solubility [5], as described previously in the Osborne classification scheme [19]. Proteins contain both hydrophobic and hydrophilic amino acids. The majority of hydrophobic moieties are buried inside the quaternary or tertiary structure due to a hydrophobic effect, and the majority of hydrophilic moieties are on the surface, free to participate in protein-water interactions. 'Salting-in' of proteins typically occurs at low salt levels, where the ions act to increase order of the protein's hydration layers and promote protein-water interactions [78–83]. However, at high levels of salt, hydration layers can be disrupted as ion-water interactions become favored over protein-water interactions in a 'salting-out' process [78–83]. As the ions attract water molecules away from the surface of the proteins, protein-protein aggregation is favored due to hydrophobic interactions. Aggregates continue to grow in size and number until they fall out of solution as a precipitate. The ability of ions to 'salt-in' or 'salt-out' proteins depends on both the ionic strength and type of cations and/or anions present, as described according to the Hofmeister series [Anions: SO_4^{2-} > HPO_4^{2-}> acetate$^-$ > Cl^- > NO_3^-; Cations: $N(CH_3)_4^+$> NH_4^+> Na^+ = K^+ > Li^+ > Mg^{2+}] [84].

Salts formed between cations and anions with higher precipitation ability in the series decrease the solubility of non-polar amino acids, favoring hydrophobic interactions to 'salt-out' proteins. On the contrary, salts formed between cations and anions with lower precipitation ability in the series weaken the hydrophobic interactions and result in increasing solubility of non-polar amino acids, thus favoring the 'salting-in' process [85]. Broadly speaking, ammonium sulfate $(NH_4)_2SO_4$ and sodium chloride (NaCl) are the most commonly used salts for research purposes [16, 86–88]. Typically in the salt extraction procedure, proteins are initially dissolved in an aqueous NaCl solution (0.3–0.5 M) [86, 88] at neutral pH, followed by a clarification procedure to remove insoluble material. Precipitation of the protein can be triggered by either diluting the supernatant with water to lower the ionic strength or by dialysis to remove the salts, resulting in the formation of protein micelles which grow in size and number until precipitation ensues. Alsohaimy et al. [87] prepared protein isolates from chickpea, lupin and lentil using IEP and ammonium sulfate precipitation. For all of these legumes, the latter method resulted in higher protein content (chickpea – 90.6%, lupin – 92.6%

and lentil – 93.0%) in comparison to the former method (chickpea – 81.4%, lupin – 87.3% and lentil – 80.0%). On the contrary, Can Karaca et al. [16] produced isolates from chickpea, faba bean, pea and lentil using IEP and a salt extraction method and found that the protein levels obtained using the IEP method (chickpea – 85.4%, faba – 84.1%, pea – 88.8%, and lentil – 81.9%) were found to be higher than the ones produced by the salt extraction method (chickpea – 81.6%, faba – 82.0%, pea – 81.1%, and lentil- – 74.7%) [16].

4. Functional properties of pulse proteins

Protein flours, concentrates and isolates can be incorporated into various foods to increase their nutritional value and/or to provide specific and desirable functional attributes [5]. These functional attributes may include solubility, gelation, emulsifying ability, oil and water absorption capacity, and foaming. Moreover, functional properties of legume proteins contribute an important aspect in determining the competitiveness of the protein ingredient or the product in the market, as they can impact the sensory, physical and chemical properties of a food, which includes texture and organoleptic characteristics. In the literature, the functional attributes of legume proteins vary considerably due to differences in the raw material, processing, extraction methods and environmental conditions used during testing.

4.1. Solubility

Protein solubility plays a major role in various food applications as a number of functional properties such as foaming, gelation or thickening, and emulsification are closely related and often dependent on protein solubility. High protein solubility may be helpful in producing food products such as beverages, infant milk powder, imitation milk and other products which require instant solubility with no residues left. For instance, imitation milk produced using lentil protein isolate was reported to have the same quality as compared to milk prepared from soy protein isolate, however had a lower quality than when pea protein isolate was used [21]. The solubility of protein depends on various attributes including hydrophobic/hydrophilic balance of the protein molecule (mainly the surface composition: polar/non polar amino acids), pI, pH, temperature, ionic strength and the type of ions present in the solution [63]. Proteins exhibit minimum solubility at their pI because of a zero net surface charge, resulting in aggregation of protein molecules into larger structures, followed by precipitation. On the contrary, when the pH values are greater or less than the protein's pI, proteins exert a positive or negative net charge into solution, repelling one another to maximize solubility.

The solubility profiles of concentrates and isolates from various pulses obtained by IEP or UF were found to be lowest between pH 4 and 6, and significantly increased with pH shifting to either more acidic or alkaline conditions [63]. Boye et al. [66] reported that the solubility of pea, chickpea and lentil protein concentrates, which were processed using IEP and UF/DF techniques, were highest at pHs 1–3 and pHs 7–10. Moreover, the solubility profile varied with different varieties where, UF-yellow pea and UF-red lentil concentrates had the highest solubility at neutral pH, while at pH 3 and 8–10 solubility was highest for only UF-red lentil.

In both cases, the lowest solubility was found for UF-chickpea (desi). The study by Can Karaca et al. [16] on five different legumes (pea, chickpea, faba bean, lentil and soybean) showed higher overall solubility (determined at neutral pH) of these legume isolates prepared by the IEP method (85.9%) as compared to ones prepared by a salt extraction method (61.5%). For the IEP method, the pea protein isolate had the lowest solubility (61.4%); soybean isolates had the highest solubility (96.5%); and pea, lentil and chickpea isolates exhibited intermediate solubility (>90.0%). However, highly variable results were obtained for the solubility of salt-extracted isolates with values of 30.1% and 96.6% for chickpea and soybean respectively, while intermediate solubility was observed for lentil (89.8%), pea (38.1%), and faba bean (52.5%). Solubility profile of isolates produced from kabuli (PBG-1, PDG-4, PDG-3, GL769 and GPF-2) and desi chickpea cultivars (L550) were found to be non-significant as a function of genotype ($p>0.05$) [89]. However, in the study of Barac et al. [38], the solubility profile of six pea genotypes (Maja, Calvedon, Miracle, L1, L2 and L3) were found to be significantly different from each other except L2 and Maja ($p<0.05$).

4.2. Oil holding and water hydration capacities (OHC, WHC)

OHC and WHC refer to the extent to which oil and water, respectively, can be bound per gram of the protein material or legume flour [5, 63]. These properties are essential with respect to maintaining the quality of a product, its shelf life and consumer acceptability (texture and mouth feel). The ability of a protein to bind oil and water is important in preventing cook loss or leakage from the product during processing or storage [63]. Failure of a protein to bind water could lead to brittle and dry characteristics of the product [5]. WHC values for pulse protein concentrates, such as pea, faba bean, lentil and chickpea, have been determined by various groups [66, 89, 90] and fall in the range of 0.6–4.9 g/g, suggesting that both pulse genotype and manner of processing could impact values. For instance, Kaur and Singh [89] found that protein isolates prepared by kabuli chickpea cultivars (PBG-1, PDG-4, PDG-3, GL769 and GPF-2) produced significantly lower WHC than desi chickpea (L550) ($p<0.05$) which clearly indicates the impact of different cultivars in assessing functionality. Boye et al. [66] reported that for all the legumes studied (red and green lentil, desi and kabuli chickpea, yellow pea), IEP protein concentrates had higher WHCs than did ones prepared by UF (with the exception of red lentil protein concentrates) although no substantial differences were observed between WHC values between the processing treatments. The yellow pea concentrate (IEP) had the highest WHC value which was much higher than those of the kabuli and desi chickpea concentrates (IEP and UF) indicating the more significant effect of pulse type compared to extraction method on WHC.

OHC values reported by various authors [86, 89, 90] for different pulses range from 1.0–3.96 g/g, and seem to depend again on the type and variety of pulse used, and the method of preparation of the protein product. Boye et al. [66] studied the UF and IEP concentrates produced from red and green lentil, yellow pea and kabuli and desi chickpea. They reported that pulse variety and processing conditions had a larger impact on the OHC of yellow pea, kabuli chickpea and red lentil concentrates as compared to those made from desi chickpea and green lentil. Moreover, UF concentrates made from yellow pea, red lentil and kabuli chickpea had significantly higher OHC than their corresponding IEP concentrates. Red lentil and yellow pea concentrates produced by UF had the highest OHC of 2.26 g/g and 1.17 g/g respectively. However, no significant differences in OHC were observed between the IEP produced concentrates ($p>0.05$) [66]. In the study of Kaur and Singh [89], chickpea protein isolates were reported to have higher OHC than the corresponding flour samples. Moreover, in contrast to WHC, the OHC of kabuli chickpea was reported to be significantly higher than desi cultivars ($p<0.05$).

The water and oil holding properties of legume proteins may be essential in formulation of food products such as meat, pasta, cookies, etc. In producing low fat meat products, water is added to substitute the fat loss. And, water holding compounds are added to prevent cooking losses and meat shrinkage which includes proteins (whey, soy and collagen), lipids (soy lecithin) and carbohydrates (flours, starches and gums) [91]. For instance, soy proteins added to ground beef improves the tenderness, moisture retention, decreases cooking losses, and inhibits rancidity [92]. Deliza et al. [93] replaced meat in ground beef mixture with hydrated textured soybean protein (15 or 30%) and found that beef patties were more tender as compared to controls, although the overall flavor quality was reduced with having less beefy flavor. However, legumes (navy beans, chickpeas, mung beans and, red kidney beans) when substituted at a level of 15% in beef mince resulted in acceptable products, with chickpea preferred over other legumes [94].

4.3. Emulsification

An emulsion is a mixture of two or more immiscible liquids (usually oil and water), where one of the liquids (the dispersed phase) is mixed in to the other (the continuous phase) in the form of small spherical droplets [95]. Emulsions are generally classified into two types: oil-in-water (O/W), in which oil droplets are dispersed within an aqueous phase (e.g., milk, mayonnaise, cream and soups); or water-in-oil (W/O), in which water droplets are dispersed within an oil phase (e.g., butter and margarine). Emulsions are thermodynamically unstable and with time separate into oil and liquid layers due to collision and coalescence of droplets [95]. Stabilizers such as emulsifiers can be used to produce stable emulsions. For instance, protein as an emulsifier acts by adsorbing onto the oil-water interface to form a viscoelastic film surrounding the oil droplets. Stability is enhanced through electrostatic charge repulsion (depending on the pH), steric hindrance or increases to the continuous phase viscosity [95].

Protein emulsifiers are used worldwide because of their ability to adsorb at the droplet surface in an O/W emulsion during the process of homogenization, thereby reducing interfacial tension. The adsorbed protein molecules present at the surface act as a separating membrane preventing coalescence with the neighboring droplets [63]. To be an effective emulsifier, protein must exhibit the following properties: fast adsorption at the oil-water interface, ability to form a protective and cohesive layer around the oil droplets, and ability to unfold at the interface [96]. Various studies reported that the emulsifying ability of legume protein concentrates or isolates are dependent on the type of legume or the method (IEP/UF/salt extraction) used in their preparation. For instance, Fuhrmeister and Meuser [71] reported that a pea protein isolate prepared by an IEP method was found to have lower emulsifying ability as compared to one prepared using UF.

Emulsion activity index (EAI) refers to the area of emulsion stabilized per gram of emulsifier or protein material and expressed as m^2/g whereas emulsion stability index (ESI) refers to the measure of stability of this emulsion as a function of the time. Emulsion capacity (EC) is the amount of oil homogenized per gram of protein material and expressed as g oil/g protein whereas creaming stability (CS) is the ability of an emulsion to resist creaming and the formation of a serum layer as time passes, and measured as %. The study conducted by Can Karaca et al. [16] on different legumes (pea, chickpea, faba bean, soybean and lentil) showed that both legume source and extraction method (IEP or UF) had significant effects on emulsifying and physicochemical properties. Both EAI and ESI were significantly affected by legume source and extraction method, whereas EC was dependent on the legume source only. However, Boye et al. [66], studying the functional properties of chickpea, lentil and pea protein concentrates, concluded that IEP and UF preparation methods had little impact on emulsifying properties. Barac et al. [38] studying functional properties of six pea genotypes reported significant differences in emulsifying properties (EAI and ESI) as a function of Genotype and pH. The EAI of pea genotypes tested in this study was significantly higher than the commercial pea protein isolates tested.

Emulsifying and other functional properties of proteins can also be improved with protein modifications such as limited enzymatic hydrolysis using proteases (e.g. trypsin). The hydrolysis reaction results in partial unraveling of protein molecules thus exposing more ionic and hydrophobic groups for interaction with oil droplets [97]. For instance, trypsin treated oat bran protein with a ~4–8% degree of hydrolysis (DH) had improved solubility, water holding, foaming and emulsifying properties as compared to those of native proteins [98]. On the contrary, Avramenko et al. [99] reported detrimental effects of trypsin mediated hydrolysis (DH~4–20%) of lentil protein isolates. Here, except zeta potential, all the physicochemical properties (surface hydrophobicity and interfacial tension) and emulsifying properties (emulsion activity and stability indices) were found to have lower values as compared to the unhydrolyzed lentil protein isolate. This suggests that processing conditions might have specific effects dependent on protein source.

Legume proteins play a vital role in the formulation of a number of novel foods (such as sausages, bologna, meat analogues, cakes and soups) by formation and stabilization of emulsions. Meat analogues are foods which are made from nonmeat ingredients, structurally similar to meat and may have the same texture, flavor, appearance, and chemical characteristics [100]. Some of the traditional foods such as wheat gluten, rice, mushrooms, tofu and legumes when added with flavors mimic the finished a meat products such as chicken, beef, sausage etc. [100]. Soybean protein is an important meat analogue since it has meat like texture and provides a similar amino acid profile to meat proteins [100]. Tofu is a widely consumed meat analogue made from soy, which provides a good source of protein, calcium and, iron. In general, the market for meat analogues is large and includes vegetarians, vegans, and people who do not eat meat products because of religious or cultural practices.

4.4. Foaming

Similar to emulsions, foams also have two immiscible phases (aqueous and gas), and require an energy input to facilitate their formation. Foams are comprised of a dispersed gas phase within a continuous aqueous phase [96]. Proteins in solution adsorb to the gas-liquid interface in a similar manner as in emulsions to form a viscoelastic film surrounding the gas bubbles that helps resist rupturing and bubble fusion [63]. In contrast to emulsions, the major driving mechanism associated with foam instability is associated with Oswald ripening, which involves the diffusion of small gas bubbles through the continuous phase in order to become absorbed into a larger gas bubble [96]. Rupture of the viscoelastic film leads to drainage of the continuous liquid phase through the film matrix. Various food products are available which use protein as a stabilizer including meringues, whipped desserts, mousses and leavened bakery products [101]. Vose [69] reported that the foaming properties of faba bean and yellow pea isolates, prepared using UF, were higher than that of skim milk powder, wheat flour and soy protein isolates. A faba bean isolate was observed to have better foaming properties than pea protein isolate.

Foaming capacity (FC) refers to the volume of foam generated after homogenization of a certain amount of protein solution whereas foam stability (FS) refers to the ability to retain foam structure and resistance in the formation of serum layer as a function of time. In the study of Sathe and Salunkhe [102] on great northern bean (*Phaseolus vulgaris* L.) protein materials, the FCs were in the following decreasing order: albumins (180%) > protein concentrate (164%) > globulins (140%) ~ egg albumin (140%) > flour (132%) > isolate (106%), where egg albumin was the standard for measuring foaming capacity. These results indicated that all great northern bean protein materials except the isolate, had FCs that were comparable to or higher than that of egg albumin. However, the foaming stabilities were as good as egg albumin, and hence the overall foaming ability was given only a fair mark [5, 102]. Boye et al. [66] studied and compared the functional properties of yellow pea, green and red lentil, and kabuli and desi chickpea protein concentrates prepared using IEP and UF techniques. In their studies, they found that foaming capacity (which ranged from 98% to 106%) was similar for pea and lentil protein concentrates irrespective of extraction method used. However, the desi and kabuli chickpea concentrates prepared by the IEP method showed higher foaming capacity than the others. In general, it was observed that chickpea showed higher foaming capacity and expansion but lower foam stability as compared to the other sources. Furthermore, variability was observed in foaming stability with kabuli and desi chickpea and green lentil concentrates prepared by the IEP method having higher foam stability values compared to concentrates prepared by the UF method. Barac et al. [38], studying the functional properties of isolates produced from six pea genotypes using the IEP method, reported significant differences in their foaming properties as a function of genotype and regardless of changes in pH. Generally, a low foam stability was observed probably because of the low concentration of protein used in the formation of the protein solution. However, foaming capacity was highest for Maja cultivar, which was significantly higher than the commercial pea protein isolate.

5. Applications

Nowadays, there has been a growing interest by the food industry towards utilizing pulse proteins in novel products due to their nutritional value, availability, low cost, desired functional properties and beneficial health effects [3]. Pulse protein concentrates and isolates are being applied in many food products such as beverages, imitation milk, baby foods, bakery products, meat analogs, cereals, snack foods, bars, and nutrition supplements. Examples of some of the food applications of pulse proteins from literature offering opportunities for novel product development are presented in **Table 1**. Pulse proteins are also used in non-food applications such as microencapsulation of bioactive ingredients. Pulse proteins can serve as good encapsulating agents due to their amphiphilic nature, ability to stabilize oil-in-water emulsions and film forming abilities. Some of the current examples of pulse protein-based microcapsules include: alpha-tocopherol [103], polyunsaturated fatty acids-rich oil [104] and conjugated linoleic acid [105] encapsulated with pea protein, flaxseed oil encapsulated with chickpea or lentil protein [106], *Bifidobacterium adolescentis* [107] and folate [108] encapsulated with chickpea protein.

Pulse protein	Application	Protein Conc'n (%)	Outcome	References
Chickpea	Pasta	5–15	Quality characteristics of the cooked pasta were not affected by increasing protein content.	[109]
Chickpea, faba bean, lentil, mung bean, smooth pea, pea, and winged bean	Bean curd	2.3–3	Chickpea and faba beans had comparable textural properties to soybean.	[110]
Lentil and white bean	Cake	3	Lentil and white bean protein extracts tested were found to be suitable to replace soy and pea in bakery products.	[111]
Pea protein	Gluten-free bread	1–6	Pea protein addition improved rheological and structural properties of the dough.	[112]
Lupin	Bread	5–10	Lupin protein addition increased the dough development time, stability and the resistance to deformation and the extensibility of the dough.	[113]
Lupin	Fermented sausage	2	Products containing lupin protein showed no difference in firmness, appearance and color compared to control.	[114]
Pea and sweet lupin (cross-linked)	Sausage-like vegetarian substitute	9	Sensory profile and textural properties were overall accepted.	[115]

Table 1. Some examples of food applications of pulse proteins.

6. Challenges for pulse protein ingredients

Application of pulse protein ingredients in food products is limited due to the formation of a green or beany off-flavor during storage [116]. The most potent odor-active volatiles have been identified in soy protein. One of the key off-flavors in soy protein is reported to be *n*-hexanal, which is a degradation product of linoleic acid. Fermentation with *Lactobacillus* or *Streptococci* strains was suggested to overcome this hurdle [117]. In the case of pulse proteins, Murat et al. [118] showed that the flavor profile is evolving during the extraction process from pea flour to pea protein extract. The odor active compounds were found to be different between pea flour and pea protein powder. Schindler et al. [116] identified 23 highly odor-active compounds in pea protein extracts including *n*-hexanal, 1-pyrroline, dimethyl trisulfide, 1-octen-3-one, 2,5-dimethyl pyrazine, 3-octen-2-one, *β*-damascenone, and guaiacol. The authors suggested that lactic acid fermentation improved the aroma of pea protein extracts by decreasing the *n*-hexanal content and reducing or masking off-flavors.

Acknowledgements

Financial support for this work was provided by the Saskatchewan Ministry of Agriculture, the Western Grains Research Foundation, and the Saskatchewan Pulse Growers.

Author details

Ashish Singhal[1], Asli Can Karaca[2], Robert Tyler[1] and Michael Nickerson[1*]

*Address all correspondence to: michael.nickerson@usask.ca

1 Department of Food and Bioproduct Sciences, University of Saskatchewan, Saskatoon, Saskatchewan, Canada

2 Aromsa A.S. GOSB 700, Gebze, Kocaeli, Turkey

References

[1] Mudryj AN, Yu N, Aukema HM. Nutritional and health benefits of pulses. Applied Physiology Nutrition and Metabolism. 2014; 39: 1197–1204. DOI: 10.1139/apnm-2013-0557

[2] Aguilera Y, Diaz MF, Jimenez T, Benitez V, Herrera T, Cuadrado C, Martin-Pedrosa M, Martin-Cabrejas MA. Changes in non-nutritional factors and antioxidant activity during germination of nonconventional legumes. Journal of Agricultural and Food Chemistry. 2013; 61: 8120–8125. DOI: 10.1021/jf4022652

[3] Duranti M. Grain legume proteins and nutraceutical properties. Fitoterapia. 2006; 77: 67–82. DOI: 10.1016/j.fitote.2005.11.008

[4] Roy F, Boye J, Simpson BK. Bioactive proteins and peptides in pulse crops: Pea, chickpea and lentil. Food Research International. 2010; 43: 432–442. DOI: 10.1016/j.foodres.2009.09.002

[5] Boye J, Zare F, Pletch A. Pulse proteins: Processing, characterization, functional properties and applications in food and feed. Food Research International. 2010; 43: 414–431. DOI: 10.1016/j.foodres.2009.09.003

[6] Adebowale YA, Adeyemi IA, Oshodi AA. Variability in the physicochemical, nutritional and antinutritional attributes of six Mucuna species. Food Chemistry. 2005; 89: 37–48. DOI: 10.1016/j.foodchem.2004.01.084

[7] Hajos G, Osagie AU. Technical and biotechnological modifications of antinutritional factors in legumes and oilseeds. Proceedings of the Fourth International Workshop on Antinutritional Factors in Legume Seeds and Oilseeds. 2004; 293–305.

[8] Vernaza MG, Dia VP, Gonzaalez de Mejia E, Chang YK. Antioxidant and anti-inflammatory properties of germinated and hydrolysed Brazilian soybeans flours. Food Chemistry. 2012; 134: 2217–2225. DOI: 10.1016/j.foodchem.2012.04.037

[9] Martin-Cabrejas MA, Diaz MF, Aguilera Y, Benitez V, Molla E, Esteban RM. Influence of germination on the soluble carbohydrates and dietary fibre fractions in non-conventional legumes. Food Chemistry. 2008; 107: 1045–1052. DOI: 10.1016/j.foodchem.2007.09.020

[10] Benitez V, Cantera S, Aguilera Y, Molla E, Esteban RM, Diaz MF, Martin-Cabrejas MA. Impact of germination on starch, dietary fibre and physicochemical properties in non-conventional legumes. Food Research International. 2013; 50: 64–69. DOI: 10.1016/j.foodres.2012.09.044

[11] Vidal-Valverde C, Frias J, Hernandez A, Martin-Alvarez PJ, Sierra I, Rodriguez C. Assessment of nutritional compounds and antinutritional factors in pea (*Pisum sativum*) seeds. Journal of the Science of Food and Agriculture. 2003; 83: 298–306. DOI: 10.1002/jsfa.1309

[12] Frias J, DiazPollan C, Hedley CL, Vidal-Valverde C. Evolution of trypsin inhibitor activity during germination of lentils. Journal of Agricultural and Food Chemistry. 1995; 43: 2231–2234. DOI: 10.1021/jf00056a049

[13] Kothekar VS, Harsulkar AM, Khandelwal AR. Low trypsin and chymotrypsin inhibitor mutants in winged bean (*Psophocorpus tetragonolobus* (L) DC). Journal of the Science of Food and Agriculture. 1996; 71: 137–140. DOI: 10.1002/(SICI)1097-0010(199605)71:1<137::AID-JSFA563>3.0.CO;2-Y

[14] Kozlowska H, Honke J, Sadowska J, Frias J, Vidal-Valverde C. Natural fermentation of lentils: influence of time, concentration and temperature on the kinetics of hydrolysis of inositol phosphates. Journal of the Science of Food and Agriculture. 1996; 71: 367–375. DOI: 10.1002/(SICI)1097-0010(199607)71:3<367::AID-JSFA591>3.0.CO;2-I

[15] Lecomte N, Zayas J, Kastner C. Soya proteins functional and sensory characteristics improved in comminuted meats. Journal of Food Science. 1993; 58: 464–466. DOI: 10.1111/j.1365-2621.1993.tb04300.x

[16] Can Karaca A, Low N, Nickerson M. Emulsifying properties of chickpea, faba bean, lentil and pea proteins produced by isoelectric precipitation and salt extraction. Food Research International. 2011; 44: 2742–2750. DOI: 10.1016/j.foodres.2011.06.012

[17] Gupta R, Dhillon S. Characterization of seed storage proteins of Lentil (*Lens culinaris* M.). Annals of Biology. 1993; 9: 71–78.

[18] Saharan K, Khetarpaul N. Protein quality traits of vegetable and field peas: Varietal differences. Plant Foods for Human Nutrition. 1994; 45: 11–22. DOI: 10.1007/BF01091225

[19] Osborne TB. The vegetable proteins. Monographs in Biochemistry. London: Longmans, Green and Co.; 1924. DOI: 10.1002/jctb.5000431704

[20] Oomah BD, Patras A, Rawson A, Singh N, Compos-Vega R. Chemistry of Pulses. In: Tiwari BK, Gowen A, Mckenna B, editors. Pulse Foods Processing: Quality & Nutritional Applications. San Diego: Academic Press; 2011. 9–55. DOI: 10.1016/B978-0-12-382018-1.00002-2

[21] Swanson BG. Pea and lentil protein extraction and functionality. Journal of the American Oil Chemists' Society. 1990; 67: 276–280. DOI: 10.1007/BF02539676

[22] Croy RR, Gatehouse JA, Tyler M, Boulter D. The purification and characterization of a third storage protein (convicilin) from the seeds of pea (*Pisum sativum* L.). Biochemical Journal. 1980; 191: 509–516. DOI: 10.1042/bj1910509

[23] Sáenz de Miera LE, Ramos J, Pérez de la Vega M. A comparative study of convicilin storage protein gene sequences in species of the tribe Vicieae. Genome. 2008; 51: 511–523. DOI: 10.1139/G08-036

[24] O'Kane FE, Happe RP, Vereijken JM, Gruppen H, van Boekel MAJS. Characterization of pea vicilin. 1. Denoting convicilin as the r-subunit of the *Pisum* vicilin family. Journal of Agricultural and Food Chemistry. 2004; 52: 3141–3148. DOI: 10.1021/jf035104i

[25] Gatehouse JA, Croy RRD, Mclntosh R, Paul C, Boulter D. Quantitative and qualitative variation in the storage proteins of material from the EEC point field bean test. In: Bond DA, editor. Vicia faba: Feeding Value, Processing and Viruses. Boston: Martinus Nuhoff Publishers; 1980. 173–190.

[26] Martensson P. Variation in legumin:vicilin ratio between and within cultivars of *Vicia faba* L. var. minor. In: Bond DA, editor. Vicia faba: Feeding Value, Processing and Viruses. Boston: Martinus Nuhoff Publishers; 1980. 159–168.

[27] Casey R, Sharman JE, Wright DJ, Bacon JR, Guldager P. Quantitative variability in Pisum seed globulins: its assessment and significance. Plant Foods for Human Nutrition. 1982; 31: 333–346. DOI: 10.1007/BF01094045

[28] Schroeder HE. Quantitative studies on the cotyledonary proteins in the genus Pisum. Journal of the Science of Food and Agriculture. 1982; 33: 623–633. DOI: 10.1002/jsfa.2740330707

[29] Murphy PA, Resurreccion AP. Varietal and environmental differences in soybean glycinin and β-conglycinin content. Journal of Agricultural and Food Chemistry. 1984; 32: 911–915. DOI: 10.1021/jf00124a052

[30] Gueguen J, Barbot J. Quantitative and qualitative variability of pea (*Pisum sativum* L.) protein composition. Journal of the Science of Food and Agriculture. 1988; 42: 209–224. DOI: 10.1002/jsfa.2740420304

[31] Kim YH, Kim SD, Hong EH. 11S and 7S globulin fractions in soybean seed and soycurd characteristics. Korean Journal of Crop Science. 1994; 39: 348–352.

[32] Cai T, Chang KC. Processing effect on soybean storage proteins and their relationship with tofu quality. Journal of the Science of Food and Agriculture. 1999; 47: 720–727. DOI: 10.1021/jf980571z

[33] Mujoo R, Trinh DT, Ng PKW. Characterization of storage proteins in different soybean varieties and their relationship to tofu yield and texture. Food Chemistry. 2003; 82: 265–273. DOI: 10.1016/S0308-8146(02)00547-2

[34] Tzitzikas EN, Vincken JP, Groot J, Gruppen H, Visser RG. Genetic variation in pea seed globulin composition. Journal of the Science of Food and Agriculture. 2006; 54: 425–433. DOI: 10.1021/jf0519008

[35] Mertens C, Dehon L, Bourgeois A, Verhaeghe-Cartryssec C, Bleckera C. Agronomical factors influencing the legumin/vicilin ratio in pea (*Pisum sativum* L.) seeds. Journal of the Science of Food and Agriculture. 2011; 92: 1591–1596. DOI: 10.1002/jsfa.4738

[36] Cousin R. Peas (*Pisum sativum* L.). Field Crops Research. 1997; 53: 111–130. DOI: 10.1016/S0378-4290(97)00026-9

[37] Utsumi S, Matsumura Y, Mori T. Structure-function relationships of soy proteins. In: Damodaran S, Paraf A, editors. Food Proteins and Their Applications. New York: Marcel Dekker, Inc., CRC Press; 1997. 257–291.

[38] Barac M, Cabrilo S, Pesic M, Stanojevic S, Zilic S, Macej O, Ristic N. Profile and functional properties of seed proteins from six pea (*Pisum sativum*) genotypes. International Journal of Molecular Sciences. 2010; 11: 4973–4990. DOI: 10.3390/ijms11124973

[39] Cserhalmi Z, Czukor B, Gajzago-Schuster I. Emulsifying properties, surface hydrophobicity and thermal denaturation of pea protein fractions. Acta Alimentaria. 1998; 27: 357–363.

[40] Rangel A, Domont GB, Pedrosa C, Ferreira ST. Functional properties of purified vicilins from cowpea (*Vigna unguiculata*) and pea (*Pisum sativum*) and cowpea protein isolate. Journal of Agricultural and Food Chemistry. 2003; 51: 5792–5797. DOI: 10.1021/jf0340052

[41] Kimura A, Fukuda T, Zhang M, Motoyama S, Maruyama N, Utsumi S. Comparison of physicochemical properties of 7S and 11S globulins from pea, fava bean, cowpea, and French bean with those of soybean - French bean 7S globulin exhibits excellent properties. Journal of Agricultural and Food Chemistry. 2008; 56: 10273–10279. DOI: 10.1021/jf801721b

[42] Koyoro H, Powers JR. Functional properties of pea globulin fractions. Cereal Chemistry. 1987; 64: 97–101.

[43] Shen L, Tang C. Emulsifying properties of vicilins: Dependence on the protein type and concentration. Food Hydrocolloids. 2014; 36: 278–286. DOI: 10.1016/j.foodhyd.2013.10.008

[44] Bora PS, Brekke CJ, Powers JR. Heat induced gelation of pea (*Pisum sativum*) mixed globulins, vicilin and legumin. Journal of Food Science. 1994; 59: 594–596. DOI: 10.1111/j.1365-2621.1994.tb05570.x

[45] Nakamura T, Utsumi S, Mori T. Mechanism of heat-induced gelation and gel properties of soybean 7S-globulin. Agricultural and Biological Chemistry. 1986; 50: 1287–1293. DOI: 10.1080/00021369.1986.10867555

[46] Danielsson CE. Seed globulins of gramineae and leguminosae. Biochemical Journal. 1949; 44: 387–400.

[47] Derbyshire E, Wright DJ, Boulter D. Legumin and vicilin, storage proteins of legume seeds. Phytochemistry. 1976; 15: 3–24. DOI: 10.1016/S0031-9422(00)89046-9

[48] Gwiazda S, Schwenke KD, Rutkowski A. Isolation and partial characterization of proteins from pea (*Pisum sativum* L.). Nahrung. 1980; 24: 939–950. DOI: 10.1002/food.19800241003

[49] Varfolomeeva EP, Danilenko AN, Bikbov TM, Grinberg VYa, Tolstogusov VB. The rheological properties of diluted solutions of 11S globulin isolated from soybeans by using selective thermal denaturation of 2S and 7S globulins. Nahrung. 1985; 28: 503–515. DOI: 10.1002/food.19860300507

[50] Mori T, Utsumi S. Purification and properties of storage proteins of broad bean. Agricultural and Biological Chemistry. 1978; 43: 577–583. DOI: 10.1080/00021369.1979.10863486

[51] Scholz G, Richter J, Manteuffel R. Studies on seed globulins from legumes. 1. Separation. Journal of Plant Physiology. 1974; 166: 163–171.

[52] Wright DJ, Boulter D. Purification and subunit structure of legumin of *Vicia faba* L. (Broad bean). Biochemical Journal. 1974; 141: 413–418.

[53] Kinsella JE. Functional properties of soy proteins. Journal of the American Oil Chemists' Society. 1979; 56: 242–258. DOI: 10.1007/BF02671468

[54] Aguilera JM, Garcia HD. Protein extraction from lupin seeds: a mathematical model. International Journal of Food Science and Technology. 1989; 24: 17–27. DOI: 10.1111/j.1365-2621.1989.tb00615.x

[55] Tyler RT, Youngs CG, Sosulski FW. Air classification of legumes. I. Separation efficiency, yield, and composition of the starch and protein fractions. Cereal Chemistry. 1981; 58: 144–148.

[56] Zheng GH, Sosulski FW, Tyler RT. Wet-milling, composition and functional properties of starch and protein isolated from buckwheat groats. Food Research International. 1998; 30: 493–50. DOI: 10.1016/S0963-9969(98)00021-0

[57] Tian S, Kyle WSA, Small DM. Pilot scale isolation of proteins from field peas (*Pisum sativum*, L.) for use as food ingredients. International Journal of Food Science and Technology. 1999; 34: 33–39. DOI: 10.1046/j.1365-2621.1999.00236.x

[58] Tyler RT. Impact milling quality of grain legumes. Journal of Food Science. 1984; 49: 925–930. DOI: 10.1111/j.1365-2621.1984.tb13243.x

[59] Sosulski F, Youngs CG. Yield and functional properties of air-classified protein and starch fractions from eight legume flours. Journal of the American Oil Chemists' Society. 1979; 56: 292–295. DOI: 10.1007/BF02671477

[60] Gueguen J. Legume seed protein extraction, processing, and end product characteristics. Plant Foods for Human Nutrition. 1983; 32: 267–303. DOI: 10.1007/BF01091191

[61] Gueguen J, Cerleti P. Proteins of some legume seeds: soybean, pea, faba bean and lupin. In: Hudson BJF, editor. New and Developing Sources of Food Proteins. London: Chapman and Hall; 1994. 145–193. DOI: 10.1007/978-1-4615-2652-0_6

[62] Pelgrom PJM, Vissers AM, Boom RM, Schutyser MAI. Dry fractionation for production of functional pea protein concentrates. Food Research International. 2013; 53: 232–239. DOI: 10.1016/j.foodres.2013.05.004

[63] Kiosseoglou V, Paraskevopoulou A. Functional and physicochemical properties of pulse proteins. In: Tiwari BK, Gowen A, Mckenna B, editors. Pulse Foods Processing: Quality & Nutritional applications. San Diego: Academic Press; 2011. 57–89. DOI: 10.1016/B978-0-12-382018-1.00003-4

[64] Schutyser MAI, Pelgrom PJM, Van der Goot AJ, Boom RM. Dry fractionation for sustainable production of functional legume protein concentrates. Trends in Food Science and Technology. 2015; 45: 1–11. DOI: 10.1016/j.tifs.2015.04.013

[65] Han Z, Hamaker BR. Partial leaching of granule-associated proteins from rice starch during alkaline extraction and subsequent gelatinization. Starch–Starke. 2002; 54: 454–460. DOI: 10.1002/1521-379X(200210)54:10<454::AID-STAR454>3.0.CO;2-M

[66] Boye JI, Akshay S, Ribéreau S, Mondor M, Farnworth E, Rajamohamed SH. Comparison of the functional properties of pea, chickpea and lentil protein concentrates processed using ultrafiltration and isoelectric precipitation techniques. Food Research International. 2010; 43: 537–546. DOI: 10.1016/j.foodres.2009.07.021

[67] Flink J, Christiansen I. The production of protein isolate from *Vicia faba*. Food Science and Technology. 1973; 6: 102–106. DOI:

[68] McCurdy S, Knipfel J. Investigation of faba bean protein recovery and application to pilot scale processing. Journal of Food Science. 1990; 55: 1093–1094. DOI: 10.1111/j.1365-2621.1990.tb01606.x

[69] Vose JR. Production and functionality of starches and protein isolates from legume seeds Field peas (*Pisum sativum* L cv. trapper) and Horse beans (*Vicia faba equina* L cv. Diana). Cereal Chemistry. 1980; 57: 406–410.

[70] Fredrikson M, Biot P, Alminger ML, Carlsson NG, Sandberg AS. Production process for high quality pea protein isolate with low content of oligosaccharides and phytate. Journal of Agricultural and Food Chemistry. 2001; 49: 1208–1212. DOI: 10.1021/jf000708x

[71] Fuhrmeister H, Meuser F. Impact of processing on functional properties of protein products from wrinkled peas. Journal of Food Engineering. 2003; 56: 119–129. DOI: 10.1016/S0260-8774(02)00241-8

[72] Singh U. Antinutritional factors of chickpea and pigeonpea and their removal by processing. Plant Foods for Human Nutrition. 1988; 38: 251–261. DOI: 10.1007/BF01092864

[73] Waggle DH, Steinke FH, Shen JL. Isolated soy proteins. In: Matthews RH, editor. Legumes: Chemistry, Technology and Human Nutrition. New York: Marcel Dekker; 1989. 99–138.

[74] Mondor M, Aksay S, Drolet H, Roufik S, Farnworth E, Boye JI. Influence of processing on composition and antinutritional factors of chickpea protein concentrates produced by isoelectric precipitation and ultrafiltration. Innovative Food Science & Emerging Technologies. 2009; 10: 342–347. DOI: 10.1016/j.ifset.2009.01.007

[75] Koros WJ, Ma, YH, Shimdzu T. Terminology for membranes and membrane processes. Pure & Applied Chemistry. 1996; 68: 1479–1489. DOI: 10.1351/pac199668071479

[76] Papalamprou EM, Doxastakis GI, Biliaderis CG, Kiosseoglou V. Influence on preparation methods on physicochemical and gelation properties of chickpea protein isolates. Food Hydrocolloids. 2009; 23: 337–343. DOI: 10.1016/j.foodhyd.2008.03.006

[77] Papalamprou EM, Doxastakis GI, Kiosseoglou V. Chickpea protein isolates obtained by wet extraction as emulsifying agents. Journal of the Science of Food and Agriculture. 2010; 90: 304–313. DOI: 10.1002/jsfa.3816

[78] Philip JC. Influence of non-electrolytes and electrolytes on the solubility of sparingly soluble gases in water the question of hydrates in solution. Journal of the Chemical Society. 1907; 91: 711–717. DOI: 10.1039/CT9079100711

[79] Glasstone S, Dimond DW, Jones EC. Solubility influences. Part II. The effect of various salts on the solubility of ethyl acetate in water. Journal of the Chemical Society. 1926; 129: 2935–2939. DOI: 10.1039/JR9262902935

[80] Glasstone S, Bridgman J, Hodgson WRP. Solubility influences. Part IV. The salting-out of aniline from aqueous solutions. Journal of the Chemical Society. 1927; 635–640. DOI: 10.1039/JR9270000635

[81] Eucken A, Hertzberg G. Aussalzeffekte und ionehydratation. Angewandte Chemie. 1950; 62: 196–196.

[82] Desnoyers JE, Jolicoeur C. Hydration effects and thermodynamics properties of ions. In: Bockris JO'M, Conway BE, editors. Modern Aspects of Electrochemistry No.5. New York: Plenum Press; 1969. DOI: 10.1007/b113773

[83] Grover PK, Ryall RL. Critical appraisal of salting-out and its implications for chemical and biological sciences. Chemical Reviews. 2005; 105: 1–10. DOI: 10.1021/cr030454p

[84] Hofmeister F. About the science of the effect of salts. Archive for Experimental Pathology and Pharmacology. 1888; 24: 247–260. DOI: 10.1007/BF01918191

[85] Nostro PLo, Ninham BW. Hofmeister phenomena: an update on ion specificity in biology, Chemical Reviews. 2012; 112: 2286–2322. DOI: 10.1021/cr200271j

[86] Paredes-Lopez O, Ordorica-Falomir C, Olivares-Vazquez MR. Chickpea protein isolates: Physicochemical, functional and nutritional characterization. Journal of Food Science. 1991; 56: 726–729. DOI: 10.1111/j.1365-2621.1991.tb05367.x

[87] Alsohaimy SA, Sitohy MR, El-Masry RA. Isolation and partial characterization of chickpea, lupine and lentil seed proteins. World Journal of Agriculture Sciences. 2007; 3: 123–129.

[88] Sun XD, Arntfield SD. Gelation properties of salt extracted pea protein induced by heat treatment. Food Research International. 2010; 43: 509–515. DOI: 10.1016/j.foodres.2009.09.039

[89] Kaur M, Singh N. Characterization of protein isolates from different Indian chickpea (*Cicer arietinum* L.) cultivars. Food Chemistry. 2007; 102: 366–374. DOI: 10.1016/j.foodchem.2006.05.029

[90] Fernandez-Quintela A, Macarulla MT, Del Barrio AS, Martinez JA. Composition and functional properties of protein isolates obtained from commercial legumes grown in

northern Spain. Plant Foods for Human Nutrition. 1997; 51: 331–342. DOI: 10.1023/A: 1007936930354

[91] Brewer MS. Reducing the fat content in ground beef without sacrificing quality: A review. Meat Science. 2012; 91: 385–395. DOI: 10.1016/j.meatsci.2012.02.024

[92] Kotula AW, Berry BW. Addition of soy proteins to meat products. Plant Proteins: Applications, Biological Effects, and Chemistry. American Chemical Society Symposium Series. 1986; 312: 74–89. DOI: 10.1021/bk-1986-0312.ch007

[93] Deliza R, Saldivar SOS, Germani R, Benassi VT, Cabral LC. The effects of coloured textured soybean protein (TSP) on sensory and physical attributes of ground beef patties. Journal of Sensory Studies. 2002; 17: 121–132. DOI: 10.1111/j.1745-459X.2002.tb00337.x

[94] Muller GG, Redden VR. Sensory and functional evaluation of ground beef patties extended with milling grade culinary beans. Proceedings of the Joint Australia and New Zealand Institute of Food Science and Technology Conference. 1995; 52: 15–19.

[95] McClements DJ. Food Emulsions: Principles, Practices, and Techniques. Boca Raton: CRC Press; 2005.

[96] Damodaran S. Protein stabilization of emulsions and foams. Journal of Food Science. 2005; 70: 54–66. DOI: 10.1111/j.1365-2621.2005.tb07150.x

[97] Panyam D, Kilara A. Enhancing the functionality of food proteins by enzymatic modification. Trends in Food Science & Technology. 1996; 7: 120–125. DOI: 10.1016/0924-2244(96)10012-1

[98] Guan X, Yao H, Chen Z, Shan L, Zhang M. Some functional properties of oat bran protein concentrate modified by trypsin. Food Chemistry. 2007; 101: 163–170. DOI: 10.1016/j.foodchem.2006.01.011

[99] Avramenko NA, Low NH, Nickerson MT. The effects of limited enzymatic hydrolysis on the physicochemical and emulsifying properties of a lentil protein isolate. Food Research International. 2013; 51: 162–169. DOI: 10.1016/j.foodres.2012.11.020

[100] Malav OP, Talukder S, Gokulakrishnan P, Chand S. Meat analogue: a review. Critical Reviews in Food Science and Nutrition. 2015; 55: 1241–1245. DOI: 10.1080/10408398.2012.689381

[101] Townsend A, Nakai S. Relationships between hydrophobicity and foaming characteristics of food proteins. Journal of Food Science. 1983; 48: 588–594. DOI: 10.1111/j.1365-2621.1983.tb10796.x

[102] Sathe SK, Salunkhe DK. Functional properties of great northern bean proteins: emulsion, foaming, viscosity & gelation properties. Journal of Food Science. 1981; 46: 71–84. DOI: 10.1111/j.1365-2621.1981.tb14533.x

[103] Pierucci AP, Andrade LR, Farina M, Pedrosa C, Rocha-Leao MHM. Comparison of alpha-tocopherol microparticles produced with different wall materials: Pea protein a

new interesting alternative. Journal of Microencapsulation. 2007; 24: 201–213. DOI: 10.1080/02652040701281167

[104] Aberkane L, Roudaut G, Saurel R. Encapsulation and oxidative stability of PUFA-rich oil microencapsulated by spray drying using pea protein and pectin. Food & Bioprocess Technology. 2014; 7: 1505–1517. DOI: 10.1007/s11947-013-1202-9

[105] Costa AMM, Nunes JC, Lima BNB, Pedrosa C, Calado V, Torres AG, Pierucci APTR. Effective stabilization of CLA by microencapsulation in pea protein. Food Chemistry. 2015; 168: 157–166. DOI: 10.1016/j.foodchem.2014.07.016

[106] Can Karaca A, Nickerson M, Low NH. Microcapsule production employing chickpea or lentil protein isolates and maltodextrin: Physicochemical properties and oxidative protection of encapsulated flaxseed oil. Food Chemistry. 2013; 139: 448–457. DOI: 10.1016/j.foodchem.2013.01.040

[107] Wang J, Korber DR, Low NH, Nickerson MT. Entrapment, survival and release of Bifidobacterium adolescentis within chickpea protein-based microcapsules. Food Research International. 2014; 55: 20–27. DOI: 10.1016/j.foodres.2013.09.018

[108] Ariyarathna IR, Karunaratne DN. Use of chickpea protein for encapsulation of folate to enhance nutritional potency and stability. Food & Bioproducts Processing. 2015; 95: 76–82. DOI: 10.1016/j.fbp.2015.04.004

[109] Yañez-Farias GA, Bernal-Aguilar V, Ramírez-Rodríguez L, Barrón-Hoyos JM. Note. Fortification of some cereal foods with a chickpea protein concentrate. Food Science & Technology International. 1999; 5: 89–93. DOI: 10.1177/108201329900500109

[110] Cai R, Klamczynska B, Baik BK. Preparation of bean curds from protein fractions of six legumes. Journal of Agricultural and Food Chemistry. 2001; 49: 3068–3073. DOI: 10.1021/jf0013398

[111] Bildstein M, Lohmann M, Hennigs C, Krause A, Hilz H. An enzyme-based extraction process for the purification and enrichment of vegetable proteins to be applied in bakery products. European Food Research & Technology. 2008; 228: 177–186. DOI: 10.1007/s00217-008-0921-z

[112] Mariotti M, Lucisano M, Ambrogina Pagani M, Ng PKW. The role of corn starch, amaranth flour, pea isolate, and Psyllium flour on the rheological properties and the ultrastructure of gluten-free doughs. Food Research International. 2009; 42: 963–975. DOI: 10.1016/j.foodres.2009.04.017

[113] Paraskevopoulou A, Provatidou E, Tsotsiou D, Kiosseoglou V. Dough rheology and baking performance of wheat flour-lupin protein isolate blends. Food Research International. 2010; 43: 1009–1016. DOI: 10.1016/j.foodres.2010.01.010

[114] Papavergou EJ, Bloukas JG, Doxastakis G. Effect of lupin seed proteins on quality characteristics of fermented sausages. Meat Science. 1999; 52: 421–427. DOI: 10.1016/S0309-1740(99)00025-X

[115] Schafer C, Neidhart S, Carle R. Application and sensory evaluation of enzymatically texturised vegetable proteins in food models. European Food Research & Technology. 2011; 232: 1043–1056. DOI: 10.1007/s00217-011-1474-0

[116] Schindler S, Zelena K, Krings U, Bez J, Eisner P, Berger RG. Improvement of the aroma of pea (*Pisum sativum*) protein extracts by lactic acid fermentation. Food Biotechnology. 2012; 26: 58–74. DOI: 10.1080/08905436.2011.645939

[117] Blagden TD, Gilliland SE. Reduction of levels of volatile components associated with the "beany" flavor in soymilk by lactobacilli and streptococci. Journal of Food Science. 2005; 70: M186–189. DOI: 10.1111/j.1365-2621.2005.tb07148.x

[118] Murat C, Bard MH, Dhalleine C, Cayot N. Characterisation of odour active compounds along extraction process from pea flour to pea protein extract. Food Research International. 2013; 53: 31–41. DOI: 10.1016/j.foodres.2013.03.049

Chapter 4

Genetic Control of Cadmium Concentration in Soybean Seeds

Souframanien Jegadeesan , Richard Ying Yu and Kangfu Yu

Additional information is available at the end of the chapter

http://dx.doi.org/10.5772/64911

Abstract

Cadmium (Cd) is a chemical element present in the soil. At high concentrations Cd can cause physiological and morphological damages to plants and it is highly toxic to human beings. Minimizing the intake of Cd and other heavy metals from food consumption is an important health issue. Efforts have been made to identify genetic elements that are involved in Cd detoxification in plants. Heavy metal transporter 3 (HMA3) plays a role in sequestration of Cd into vacuoles in soybean (*Glycine max*). Inheritance studies revealed that low Cd accumulation in soybean seed is controlled by a major gene (*Cda*1) with the allele for low accumulation being dominant. Major QTL for seed Cd accumulation, *Cda*1 and *cd*1, have been identified independently for low Cd accumulation and both mapped to the same location as on LG-K (Chromosome 9) with simple sequence repeat (SSR) markers. A single nucleotide substitution causing a loss of function of the ATPase was found. The SSR markers linked to the *Cda*1 and *Cd*1gene(s)/or QTLs and the SNP marker in the P1B-ATPase metal ion transporter gene in soybean can be utilized in marker assisted selection (MAS) for developing food grade soybean varieties.

Keywords: cadmium, soybean, SSR markers, QTLs, marker-assisted selection

1. Introduction

Cadmium (Cd) is a highly toxic element for human beings because of its extremely long biological half-life. Vast areas of agricultural soils are contaminated with Cd through the use of super phosphate fertilizers, sewage sludge, and inputs from the mining and smelting industries [1]. Cd^{2+} is readily taken up by roots and can be translocated into aerial organs, where it affects photosynthesis and consequently root and shoot growth. At high concentrations, Cd can cause severe physiological and morphological damages to plants, such as stunted root and shoot growth [2–4], chlorosis, decreased reproducibility [5], and reduced water and nutrient uptake [6]. Cd stress can affect enzyme activities [3, 7], alter membrane permeability [6], and disrupt cell transport processes [8]. Cd stress can also disturb cellular redox control [9], damage the light-harvesting complex II [10] and photosystems I and II [11], and decrease carbon assimilation and chlorophyll content [12]. Soybean has long been a staple food for Asians, especially as soymilk, tofu, and oil [13]. Many soybean cultivars can accumulate high Cd concentration in seed when grown on Cd-polluted soil [2, 14].

Cd can accumulate in the human body over time from ingestion of food containing Cd, leading to a risk of chronic toxicity with excessive intake. In humans, it can damage kidneys, causing a loss of calcium and associated osteoporosis [15]. It is desirable to limit the concentration of Cd in crops used for human consumption to reduce potential health risks. Due to growing concern about safety of foods and human health, Codex Alimentarius Commission of Food and Agriculture Organization/World Health Organization (FAO/WHO) has proposed an upper limit of 0.2 mg kg^{-1} for Cd concentration in soybean grain [16]. The results of a large-scale survey of domestic agricultural products revealed that the Cd concentration of 16.7% of soybean seeds exceeded the international allowable limit of 0.2 mg kg^{-1} proposed by the Codex Committee until 2001, which is much higher than that of other upland crops [17].

Considering the health issues due to the intake of Cd and other heavy metals through food grains, cultivars with reduced uptake of these metals are needed for human health. Breeding cultivar with reduced Cd is an attractive method for changing the element profile of crops as the benefit will persist in the seed that can reduce the requirement for other management practices [18]. The amount of Cd that enters the human diet from a crop depends on the amount of Cd accumulated in the portion of the plant that is edible rather than solely on total plant uptake. Both accumulation and distribution of Cd in the plant differ depending on the species, the cultivar, and the growing conditions. In general, the distribution of Cd within the plant is influenced by transport from roots to the shoots via the xylem, transfer from the xylem to the phloem, and transport through the phloem from sources to sinks and other environmental factors [19].

2. Genetic variation for Cd uptake

Natural variation occurs in the uptake and distribution of essential and nonessential trace elements among crop species and among cultivars within species. Plant breeding can be an important tool to either increase the concentration of desirable trace elements or reduce that of potentially harmful trace elements such as Cd. Since the Cd trait is highly heritable, incorporation of the genes influencing low Cd accumulation can help to reduce the average grain Cd to levels below the recommended international limit. The allele for low Cd concentration can be incorporated into other cultivars through breeding program without affecting

http://dx.doi.org/10.5772/64911

other agronomic traits [20]. Cd uptake depends both on the Cd concentration in the soil and on the characteristics of the specific cultivars. Accumulation of large amounts of Cd in the root may limit the accumulation of Cd in edible aboveground portions of the plant. It was reported that Cd concentration in soybean seeds was reduced when high accumulating soybean lines (rootstock) were grafted with low accumulating lines. This indicated that the Cd accumulation in the seed was reduced by high accumulation in the root and was controlled by the rootstock cultivar [21]. Differences in seed Cd concentration among different varieties may be in part related to differences in the abilities of plants to control movement of Cd from the xylem into the phloem, and via the phloem to the soybean seeds [2, 22, 23]. There was also considerable genetic variation observed among soybean cultivars [2, 23–26], with low Cd cultivars appearing to retain more Cd in the root and translocate less to the seed than high Cd cultivars [22].

In field-grown soybean, a wide range of Cd concentrations varying from 0.08 to 1.1 mg kg^{-1} in seed have been reported depending on growing environment and soybean genotype [2, 27–29]. Low soil pH, vicinity to mining sites or sludge applications, has contributed to an increased Cd level in soybean seed [28–30]. In most studies, soybean Cd levels were considerably higher in roots, stems, leaves, or pods than in seeds. Moreover, a high soil Cd concentration is also toxic to soybean reducing plant growth and photosynthesis apart from other effects [31]. Due to genetic differences in soybean cultivar for seed Cd accumulation, a three- to sixfold Cd concentration increase was observed between lowest and highest accumulating genotypes. It was reported that the variation in the Cd accumulation level between genotypes was due to differences in both uptake and Cd retention of the roots [2]. Cadmium concentration in roots showed far higher than that in shoots of soybean genotypes. The root morphological traits such as the total root length (RL), root surface area (SA), and root volume (RV) were closely related to Cd tolerance at young seedlings under Cd treatments [26].

Genotypic differences in Cd uptake and distribution were observed in soybeans cultivated in pot and under low Cd concentrations in the field [2]. Cultivars with low Cd uptake accumulated much higher Cd in their roots than those of the cultivar with high Cd uptake [32]. Decreasing soil Cd concentration reduced Cd concentration in soybean seeds [33]. Interaction of Cd and nitrogen resulted in decreased Cd uptake by soybean seedling roots cultivated at a high nitrogen nutrition level [34]. Cd adversely affected soybean growth, nodulation, and N_2 fixation as a function of time and increase in Cd concentration [35]. The risk of toxicity from Cd in food is influenced not only by Cd concentration but also by concentrations of other trace elements such as Zn and iron (Fe) [36]. Breeding programs are underway to increase the concentration of essential trace elements to enhance the nutritional value of staple crops. Breeding programs to increase concentrations of essential trace elements would have the combined benefit of enhancing the nutritional value of staple crops while reducing the bioavailability of Cd, particularly if low Cd was included as a selection criterion [20].

Growth stage or the age of the plants and the time of exposure to the heavy metal also affect Cd absorption and distribution between different cultivars and between plant parts. The soybean cultivar "Doko" showed an increase in Cd concentration in the roots from the VC (cotyledon stage) to V2 (second node) stage while the cultivar "Bossier" showed the opposite trend in roots. The Cd content of both cultivars (cvs) in stems, however, did not change much increase the Cd concentration in both low and high Cd cultivars. Cultivar selection can be effective in reducing the potential Cd concentration in crops. However, the availability of an inexpensive methods to detect and select for genetic differences in Cd concentration at an early developmental stage will reduce the time and cost of a breeding program [23, 51, 56, 57].

5. Marker-assisted selection for low Cd accumulation in soybean

5.1. Developing markers for marker-assisted selection of low Cd accumulation

Marker-assisted selection (MAS), the use of molecular markers linked to or located at a desired gene locus, could be an alternative to phenotypic selection. In soybean, DNA markers linked to low Cd accumulation were identified using recombinant inbred line population ($F_{6:8}$) derived from the cross AC Hime (high Cd accumulation in seeds) and Westag-97 (low Cd accumulation in seeds). The distribution of Cd concentration of 166 RILs ranged from 0.067 to 0.898 mg kg^{-1}, with a mean of 0.268 ± 0.013 mg kg^{-1}. Of the166 RILs analyzed, 87 had ≤0.2 mg kg^{-1} and 79 had ≥0.21 mg kg^{-1} (**Figure 1**). Using the RIL population, seven simple sequence repeat (SSR) markers, SatK138, SatK139, SatK140 (0.5 cM), SatK147, SacK149, SaatK150, and SattK152 (0.3 cM), were reported to be linked to *Cda1* in soybean seed (**Table 1**). It was also reported that all the linked markers were mapped to the same linkage group (LG) K, indicating that a major gene affecting Cd accumulation could be located in the region (**Figure 2**).

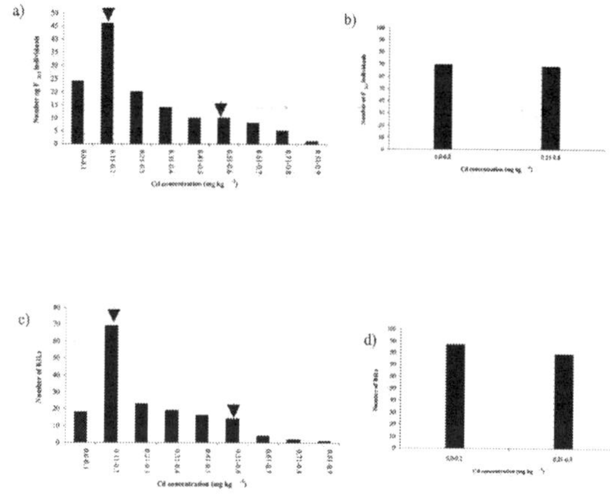

Figure 1. Frequency distribution of seed cadmium concentration in the AC Hime × Westag-97 population. (a) Frequency distribution in the $F_{2:3}$ population in 2005. (b) Lines with low Cd (≤0.2 mg kg^{-1}) and high Cd concentration (≥0.2 mg kg^{-1}) in the $F_{2:3}$ population. (c) Frequency distribution in $F_{6:8}$ RIL population in 2008. (d) Lines with low Cd (≤0.2 mg kg^{-1}) and high Cd concentration (≥0.2 mg kg^{-1}) in the $F_{6:8}$ RIL population. The arrow indicates the level of the parental lines. Solid and dashed arrows indicate the AC Hime and the Westag-97 parent, respectively.

Primer	Primer sequence (5′–3′)	Repeat DNA	Size (bp)	Reference
SatK 138F	AATGAATGTGATGTGATTTGTCA	$(AT)_{29}$	313	Jegadeesan et al. [53]
SatK 138R	TGAGTTAGGTAAGATGGTCATTAAAA			
SatK 139F	AACTAAACAATGTAATGTGATTTGTCA	$(AT)_{25}$	201	
SatK 139R	AAGTTAAACCTTAATTCAAGAAATGTG			
SatK 140F	AACTTTAATCGAAAAGTTATTGCTGA	$(AT)_{13}$	200	
SatK 140R	CAGCTAGAACCTAGAAGATTACGC			
SatK 147F	CCATGGATATCTCCTAATCTCCTG	$(AT)_{18}$	203	
SatK 147R	TCTGCAAATTAAAACTTAGAGGGTG			
SacK 149F	TGAACACATGCTCAACTTGTCA	$(AC)_{18}$	236	
SacK 149R	CGTGTGGTTGCTATTAACTAAATGA			
SaatK 150F	TGATGTCTCCGTACATAAAAGATCAC	$(AAT)_8$	286	
SaatK150R	CTTCAACCATACGCTTGTGAA			
SattK 152F	AAAATGTGACCAAACGGGAC	$(ATT)_{20}$	205	
SattK 152R	CACGCCAGTAAATCAAAACTCA			
Gm09: 4770663-F	AAAGCACGGCTGCTTATATAGTT			Benitez et al. [27]
Gm09: 4770663-F	CGTCGTGCATGTGTTATATATTATT			
Gm09:4790483-F	AAGCCCACGATTAGTACTTGGA			
Gm09:4790483-R	ACCAGGCATGTAGTTTCTGTAGC			
Gm-dCAPS-HMA1-F	TGACATCGGTATCTCACTGG		90	Benitez et al. [74]
Gm-dCAPS-HMA1-R	ATGACATTCTCAATTAGCTTTC			
GmHMA3w-F	GCTGACATCGGTATCTCA			Wang et al. [61] (**Figure 5**)
GmHMA3w-R	GCATTGCCTGTTTCATTTG			

Table 1. DNA markers linked to low cadmium accumulating locus Cda1 and Cd1 located on soybean linkage group K (Gm:09).

The closest flanking SSR markers linked to *Cda1* were validated using diverse soybean cultivars and a parallel population (RILs) involving Leo 9 × Westag-97. SSR markers SatK147, SacK149, and SattK152 clearly differentiated the high and low Cd-accumulating genotypes tested in soybean [53]. In order to identify QTL affecting Cd accumulation, a linkage map constructed with 161 markers identified a major QTL associated with low Cd concentration in the soybean seeds. The QTL for low Cd accumulation was also reported to be mapped on the same location as *Cda*1 on LG-K, and accounted for 57.3% of the phenotypic variation [53]. SSR markers closely linked to *Cda*1 in soybean seeds have the potential to be used for MAS to develop low Cd-accumulating cultivars in a breeding program. In a similar mapping approach, Benitez et al. [27] reported a major QTL cd1 affecting seed Cd content using RILs derived from a cross between two cultivars: Harosoy (with high seed Cd content) and Fukuyutaka (with low Cd content). This major QTL, cd1, was identified on chromosome 9 (LG-K) across years and generations which accounted for 82, 57, and 75% of the genetic variation. Near isogenic lines (NILs) were used to confirm the effect of the QTL and the peak of the QTL that was located in the vicinity of two SSR markers, Gm09:4770663 and Gm09:4790483 (**Table 1**). The separate studies revealed a major QTL for seed *Cd* content, *Cda*1 at a similar genomic location, suggesting that *cd*1 and *Cda1* may be identical.

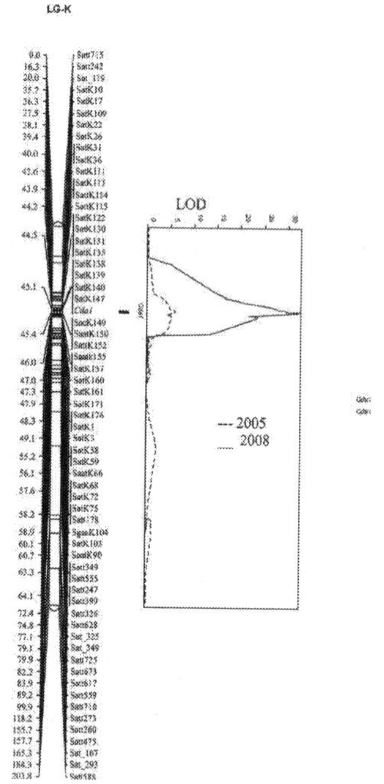

Figure 2. Linkage group-K which corresponds to chromosome 9 (Gm: 09) indicating the location of the newly developed SSR markers and the location of the major gene Cda1 or QTL controlling low Cd accumulation in soybean seed. Location of the major QTL associated with low Cd accumulation mapped on the LG-K with its LOD score values are shown for the AC Hime × Westag-97 $F_{2:3}$ (2005) and $F_{6:8}$ RIL (2008) populations.

5.2. Validation of markers linked to low Cd accumulation

SSR markers linked to low Cd accumulation were validated using diverse soybean genotypes differing in their seed Cd concentration. Of the 12 primers evaluated, three (SatK 147, SatK 149, and SattK 152) effectively differentiated all the high and low Cd genotypes and could be used effectively in MAS for identifying low Cd-accumulating genotype in soybean seed [53]. The reliability of these linked SSR markers was also tested using another RIL population (95 lines) involving Leo 9 × Westag-97. Leo 9 and Westag-97 had seed Cd concentrations of 0.435

± 0.046 and 0.170 ± 0.001 mg kg^{-1}, respectively. The concentration of Cd in the seeds of the Leo × Westag-97 population varied from 0.065 to 0.878 mg kg^{-1}, with a mean of 0.305 ± 0.019 mg kg^{-1}. Of the 95 lines analyzed, 42 were in the low (≤0.2 mg kg^{-1}) and 53 were in the high (≥0.21 mg kg^{-1}) category. Eight SSR primer pairs (SatK 122, SatK 131, SatK 140, SatK 147, SatK 149, SaatK 150, SattK 152, and SaatK 155) were found to be linked to the *Cda*1 gene [53]. It was found that the relative positions of the markers were found to be the same as was found in the AC Hime × Westag-97 population with minor variation in the distances, which often occurs with different mapping populations [58].

Furthermore, these SSR markers were validated for their suitability to discriminate the low and high Cd-accumulating soybean genotypes grown in Europe [59]. The reliability of the SSR markers for the *Cda1* gene revealed that more than half (12) of the examined soybean cultivars carried the allele for low Cd accumulation. The SSR analysis identified soybean cultivars with potential health risk when grown in metal-polluted areas, regardless of their natural tolerance [59]. Vollmann et al. [60] validated the low seed Cd accumulation trait based on the *Cda*1 locus and the associated Sack149 marker. Out of 48 genotypes evaluated, 19 exhibited the allele associated with low and 29 with high Cd accumulation in the seed. SSR marker Sack149 amplified a single polymerase chain reaction (PCR) product was visible in each of the accessions, and no other alleles than the two described for the SacK149 marker were found in any of the genotypes analyzed. Sack149 marker is clearly effective over a range of different genotypes, and thus soybean lines with reduced seed Cd concentration could be selected without the need for extensive and costly field testing in locations with Cd-contaminated soils.

6. Candidate gene(s) controlling Cd accumulation in soybean

Soybean genome sequence available from phytozome (http://www.phytozome.net/soybean.php) via SoyBase (http://soybase.org/gbrowse/cgi-bin/gbrowse/gmax1.01/) was analyzed to identify the candidate genes located between the tightly linked flanking markers (SatK 140 and SaatK 155). Three potential genes homologous to serine-threonine protein kinase, plant type (nt. 4909157–4913830) and two homologous to cation-transporting ATPase (nt. 4918664–4926453 and 5011045–5020110) were identified based on the predicted gene model for the DNA sequence from nt.4909157 to nt.5020110, flanked by SatK 140 and SaatK 155. "Moreover, 13 soybean ESTs, including TA47883_3847 [plasma membrane H^+-ATPase (*Sesbania rostrata*)], TA65152_3847 [Protein kinase, (*Medicago truncatula*)], and AW152957 [Adagio-like protein 1 (*Oryza sativa*)], were also aligned to this genomic region" [53]. There are four SSR markers (SatK 147, SacK 149, SaatK 150, and SattK 152) found in the vicinity of the genes. Of these SSR markers, SattK 152 is reported to be located in the candidate gene plasma membrane H^+-ATPase [53].

In another parallel study, the evaluation of the *Cda* 1 locus and the SSR marker genotype indicated that the candidate gene should be located in the 184.3-kb genomic region between the SatK130 and SacK 149 markers [61]. According to the gene annotation in the SoyBase (http://soybase.org/) [62], six annotated genes in this region were found (**Figure 3**).

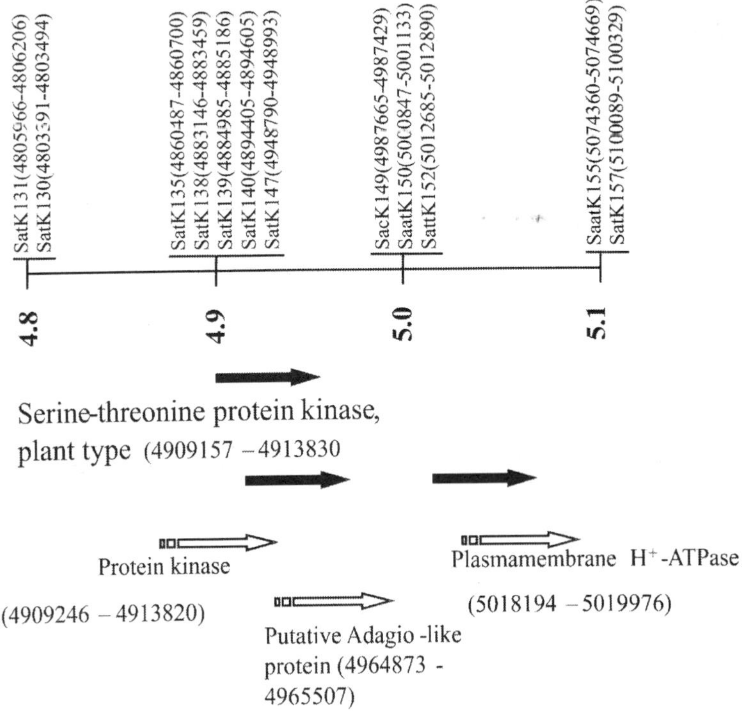

Figure 3. Physical location of the SSR markers in Gm:09 tightly linked to Cda1 controlling low Cd accumulation in soybean seeds. Putative genes located in the vicinity of the tightly linked markers (http://soybase.org/) based on the predicted and known gene function with EST support for soybean genomic sequences are shown.

Among them, Glyma09g06170 encodes a putative heavy-metal transporter (GmHMA3). Its homologs, AtHMA3 and OsHMA3, which belong to P1B-ATPases and localized on the vacuolar membrane in *Arabidopsis thaliana* and rice, were reported to have potential to sequester Cd into vacuoles to limit Cd transport to the xylem [63, 64]. On comparing the full-length cDNA sequence of GmHMA3a from AC Hime (a high Cd accumulator in seeds, GI# JN187676) and GmHMA3w from Westag 97 (a low Cd accumulator in seeds, GI# JN187675), it was found that the two gene sequences are identical and have nine exons and eight introns except for a single nucleotide polymorphism (SNP) at nucleotide position 1823 in GmHMA3a. This single nucleotide change from G to A resulted in the substitution of glutamic acid (E) with glycine (G) at position 608, which is highly conserved in AtHMA3 and OsHMA3, even in AtHMA2 and AtHMA4 [61]. HRM (high-resolution melt) analysis genotyped the SNP in AC Hime, Westag 97, and the 166 RILs; the results indicated GmHMA3w (0.3 cM away from *Cda*1) is significantly associated with low seed Cd concentration in the RILs. To validate the SNP, 20 diverse soybean cultivars were genotyped and confirmed by sequencing. It was found that the 13 high Cd accumulators had the GmHMA3a allele while the seven low Cd accumu-

http://dx.doi.org/10.5772/64911

lators had the GmHMA3w allele. GmHMA3w was found to be associated with low Cd level in soybean seeds and the SNP marker effectively differentiated high from low Cd phenotype [61]. Gene expression studies revealed that GmHMA3 expressed only in roots of AC Hime and Westag 97 (**Figure 4**), indicating that GmHMA3 plays an essential role in the transport of some divalent metals in roots [61].

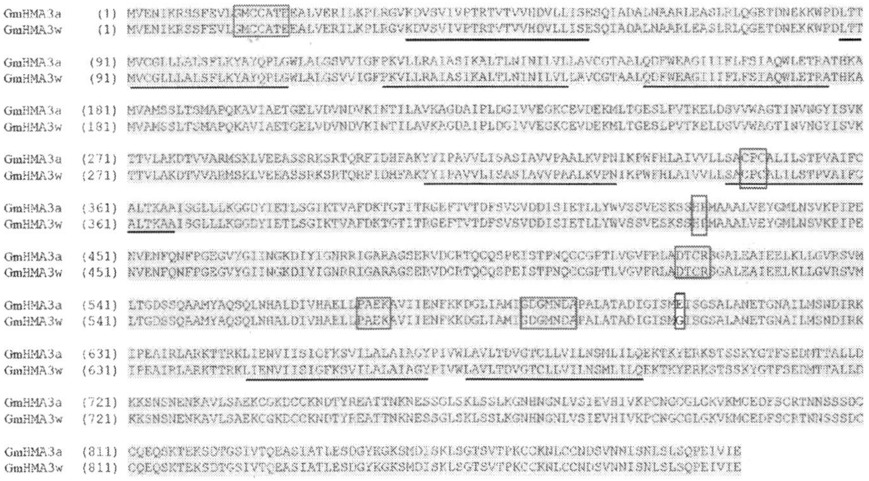

Figure 4. Alignment of two allelic amino acid sequences (GmHMA3a and GmHMA3w) from AC Hime and Westag 97, respectively. Blue box indicated the single amino acid mutation. Red boxes showed all typical motifs of P_{1B}-HMA. Transmembrane domains were underlined with black lines.

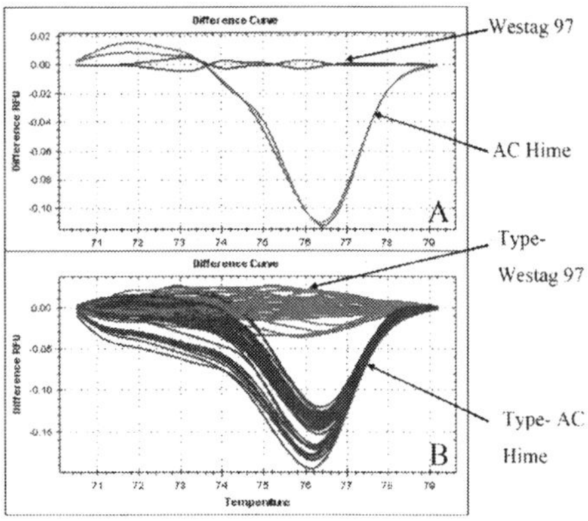

Figure 5. HRM analysis of the SNP for the parents (A) and the 166 RILs (B).

The regulation of metal homeostasis is complex and controlled by several metal-specific and metal nonspecific genes located in different membranes and long-distance transport systems to move throughout the plant. The presence of higher levels of heavy metal ions in the soil triggers a wide range of cellular responses including the synthesis of metal-detoxifying peptides and change in gene expression. Cd- and copper-responsive genes have been shown to code for signal transduction components, such as *Arabidopsis* mitogen-activated protein kinase kinase kinase (MAPKKK) MEKK1, stress-induced proteins, transcription factors, proteins participating in protein folding, and sulfur and glutathione metabolism [65–67]. MAPKs are pro-directed Ser/Thr kinases phosphorylating numerous substrates in different cellular compartments and thereby shown to involve in the signal transduction in the form of a phosphorylation cascade from upstream kinase to downstream targets. Cd ion-activated distinct mitogen-activated protein kinases were reported in alfalfa seedlings [68].

In soybean, candidate genes related to heavy metal transport or homeostasis were located in the vicinity of the identified QTL (Cda1). Protein kinase, putative adagio-like protein, and plasma membrane H^+-ATPase were found in the QTL vicinity. Genes uniquely induced by Cd ions in *Arabidopsis* showed a high percentage of genes with "kinase activity" (16.7%) [69]. In soybean, the influx of Cd across the plasma membrane of root cells has been shown to occur via a concentration-dependent process exhibiting saturable kinetics, indicative of metabolically mediated membrane transport process [70]. Cd seems to have differential-inhibiting effects on ATPase activity and proton transport activity in oat roots [71]. Evidence from previous studies suggests that protein kinases modulate the plant plasma membrane ATPase activity, and the ATPase probably contains multiple phosphorylation sites that may affect its activity in different ways [72]. The presence of protein kinase, and plasma membrane H^+-ATPase genes near the tightly linked SSR markers, suggests that the regulation of this enzyme may play a vital role in Cd stress [53]. This was later supported by a major QTL-controlling Cd concentration (*cd*1) identified in soybean [27]. Analysis of the genome sequence of Williams 82 from Sat_119 (Gm09:3585450) to Satt178 (Gm09:5438776) that flanks the *cd1* revealed the presence of P_{1B}-ATPase gene (Glyma09g06170.1, Gm09:4918664 to Gm09:4926453) in the vicinity, which had been implicated in the transport of a range of essential and also potentially toxic metals across cell membranes (e.g., Cu^{2+}, Zn^{2+}, Cd^{2+}, Pb^{2+}) [73].

The gene was designated as *GmHMA1,* and cDNA sequence analysis revealed the presence of two types of transcript candidates with different lengths (3719 and 3929 bp) that were identified for this gene and designated as *GmHMA1a* and *GmHMA1b* [74]. *GmHMA1a* sequence revealed the presence of nine exons and eight introns which are similar to the one identified in AC Hime [61]. In *GmHMA1b,* however, 210-bp nucleotides corresponding to the eighth intron were retained as part of an exon (i.e., were not spliced out), resulting in a structure with eight exons and seven introns. Hence, it was concluded for the existence of alternative splicing in the *GmHMA1*gene of soybean [74]. Evidence for such alternative splicing events of the intron-retention type in *GmHMA*1 has also been reported in soybean blue light photoreceptors (cryptochrome multigene family genes) GmCRY1b, GmCRY1c, GmCRY1d, and GmCRY2a [75]. The open-reading frame of *GmHMA*1a contained 2655 nucleotides with 1064 bp of the 3′-untranslated region. The putative polypeptide of *GmHMA1a* consisted of 885 amino acids with molecular mass and isoelectric points of 95,135, and 5.86, respectively. In *GmHMA1*b, translation was prematurely terminated, resulting in a polypeptide consisting of 559 amino acids due to an alternative splicing that generated a stop codon around the middle of the region corresponding to the eighth intron. In *GmHMA1a* one base substitution from G to A at nucleotide position 2095 resulted in changed amino acid from glycine to glutamic acid at amino acid number 608, but it did not affect the putative amino acid translation of *GmHMA1b* because alternative splicing generated a stop codon upstream of the base substitution [74]. No catalytic domains have been ascribed to the region of amino acid substitution, but it was located immediately downstream of the ATP-binding domain. The glycine residue at the site of amino acid substitution was fully conserved in *AtHMA*3, *AtHMA*4, *AtHMA*6, and *AtHMA*7 [76, 77], suggesting that *GmHMA1a* of Fukuyutaka is the wild type. Similar to *AtHMA*3 and *AtHMA*4, the expression of *GmHMA1* was substantially lower than actin and was predominant in roots compared with leaves [76, 77]. Using the SNP location, dCAPS primers were designed to produce a 95-bp fragment in Harosoy, Fukuyutaka, and the NILs. After *BmrI* digestion, a shorter band of 70-bp fragments observed in Fukuyutaka and the NIL of Fukuyutaka type was designated as Gm-dCAPSHMA1 [74]. Linkage mapping revealed that the marker (Gm-dCAPSHMA1) was assigned to a position identical to Gm09:4790483 and located around the three markers, Gm09:4770663, Gm09:4790483, and Gm-dCAPS-HMA1, spanning 0.6 cM. The genotype of Gm-dCAPS-HMA1 was significantly associated with seed Cd concentration. The genotype and Cd concentration completely co-segregated in the RILs. The presence of P1B-ATPases near the marker location suggested that it may be present at the intracellular membranes and be responsible for compartmentation of metals, for example, sequestration in the vacuole, Golgi, or endoplasmic reticulum, or they may be present at the plasma membrane and function as efflux pumps removing potentially toxic metals from the cytoplasm [73, 76, 77].

Wang et al. [78] studied gene expression pattern of the low and high Cd-accumulating soybean genotypes Westag-97 and AC Hime and reported different expression levels of five metal nonspecific genes, a receptor-like serine/threonine-protein kinase (RSTK, glyma09g06160), a plasma membrane H^+-transporting ATPase (H^+-ATPase, glyma09g06250), an iron-sulfur cluster scaffold protein nfu-related (ISCP, glyma09g06300), and two uncharacterized conserved protein (UCP1, glyma09g06220, and UCP2, glyma09g06310), which were previously found at the *Cda1* locus [53]. The responses of the five genes at the *Cda1* locus to Cd treatment were studied using soybean genotypes differing in Cd sequestration and translocation. Westag 97, a low seed Cd accumulator that sequestrates Cd in roots and restricts it from loading into xylem and transporting to leaves and seeds, and AC Hime, a high seed Cd accumulator that has a smaller capacity of Cd accumulation in roots but translocates and stores more Cd in stems and leaves, were used for gene expression studies. The transcriptional levels of the five genes in both AC Hime and Westag 97 were altered in response to the external Cd treatment [78].The expression levels of RSTK were significantly increased by Cd in AC Hime but were decreased in Westag 97. These results indicated that the RSTK is probably involved in Cd transportation. RSTK can boost Cd transporting into stems and leaves in AC Hime through elevating its expression levels and limiting Cd transporting into leaves and stems in Westag 97 through reducing its expression level. The RSTK family is involved in signal transduction pathways in plants and interacts with membrane receptor proteins. Several studies have shown that the marker along with the linked Sack149 SSR marker, which will also reduce the cost involved in the Cd analysis. The cost involved in the MAS for one sample will be approximately $1–2, compared to $10–23 for Cd analysis in an established laboratory. In conclusion, the low seed Cd accumulation trait from the *Cda1* locus and its tightly linked SSR and SNP markers were clearly effective over a range of different genotypes, and thus soybean lines with reduced seed Cd concentration could be selected without the need for extensive and costly field testing in locations with Cd-contaminated soils. In addition, there is a possibility to study further mechanisms of controlling seed Cd concentration either on the root or on the shoot level, which is inferred from significant variation in seed Cd concentration within the two marker locus classes of the *Cda*1 gene and transgressive segregation.

Transgenic experiments may be necessary to determine the function of *GmHMA*1a and to verify whether the amino acid substitution affected transport and accumulation of Cd in seeds. Considering the human health issues due to Cd accumulation, the utilization of the soybean *Cda*1 locus for the selection of genotypes low in seed Cd is particularly of importance for food-grade soybean. Due to the current expansion of soybean production to new production regions with partly unknown heavy metal concentration of soils, the cultivation of low Cd-accumulating varieties would contribute to better food safety for soy food products.

Author details

Souframanien Jegadeesan[1], Richard Ying Yu[2] and Kangfu Yu[2*]

*Address all correspondence to: kangfu.yu@agr.gc.ca

1 Nuclear Agriculture and Biotechnology Division, Bhabha Atomic Research Centre, Trombay, Mumbai, India

2 Harrow Research and Development Centre, Agriculture and Agri-Food Canada, Harrow, ON, Canada

References

[1] McGrath SP, Zhao FJ, Lombi E. Plant and rhizosphere processes involved in phytoremediation of metal-contaminated soils. Plant and Soil. 2001;232:207–214.

[2] Arao T, Ae N, Sugiyama M, Takahash M. Genotypic differences in cadmium uptake and distribution in soybeans. Plant and Soil. 2003;251:247–253.

[3] Tamás L, Bočová B, Huttová J, Mistrík I, Ollé M. Cadmium-induced inhibition of apoplastic ascorbate oxidase in barley roots. Plant Growth Regulation. 2006;48:41–49.

[4] Yeh C-M, Hsiao L-J, Huang H-J. Cadmium activates a mitogen-activated protein kinase gene and MBP kinases in rice. Plant and Cell Physiology. 2004;45:1306–1312.

[5] Larsson EH, Bornman JF, Asp H. Influence of UVB radiation and Cd^{2+} on chlorophyll fluorescence, growth and nutrient content in *Brassica napus*. Journal of Experimental Botany. 1998;323:1031–1039.

[6] Sanit`a di Toppi L, Gabbrielli R. Response to cadmium in higher plants. Environmental and Experimental Botany.1999;41:105–130.

[7] Smeets K, Ruytinx J, Semane B. et al. Cadmium-induced transcriptional and enzymatic alterations related to oxidative stress. Environmental and Experimental Botany. 2008;63:1–8.

[8] Fojtová M, Fulnečková J, Fajkus J, Kovařík A. Recovery of tobacco cells from cadmium stress is accompanied by DNA repair and increased telomerase activity. Journal of Experimental Botany. 2002;53:2151–2158.

[9] Clemens S, Schroeder JI, Degenkolb T. *Caenorhabditis elegans* expresses a functional phytochelatin synthase. European Journal of Biochemistry. 2001;268:3640–3643.

[10] Krupa Z. Cadmium-induced changes in the composition and structure of the light-harvesting complex II in radish cotyledons. Physiologia Plantarum. 1998;73:518–524.

[11] Siedlecka A, Baszynsky T. Inhibition of electron flow around phytosystem I in chloroplasts of cadmium-treated maize plants is due to cadmium-induced iron deficiency. Physiologia Plantarum. 1993;87:199–202.

[12] DalCorso G, Farinati S, Maistri S, Furini A. How plants cope with cadmium: staking all on metabolism and gene expression. Journal of Integrative Plant Biology. 2008;50:1268–1280.

[13] Tay SL, Perera CO. Physicochemical properties of 7S and 11S protein mixtures coagulated by glucono-δ-lactone. Journal of Food Science.2004;69:139–143.

[14] Sugiyama M, Ae N, Hajika M. Developing of a simple method for screening soybean seedling cadmium accumulation to select soybean genotypes with low seed cadmium. Plant and Soil. 2011;341:413–422.

[15] Kazantzis G. Cadmium, osteoporosis and calcium metabolism. Biometals. 2004;17:493–498.

[16] Codex Alimentarius Commission. Report of the 33rd Session of the Codex Committee on Food Additives and Contaminants, Hague, The Netherlands; March 12–16, 2001; p. 285.

[17] Ministry of Agriculture, Forestry, and Fisheries of Japan. Investigation of Cd concentration of staple crops: the outline of the results, Tokyo; 2002.

[18] Cupit M, Larsson O, de Meeus C, Eduljee GH, Hutton M. Assessment and management of risks arising from exposure to cadmium in fertilisers: II. Science of the Total Environment. 2002;291:189–206.

[19] Riesen O, Feller U. Redistribution of nickel, cobalt, manganese, zinc, and cadmium via the phloem in young and maturing wheat. Journal of Plant Nutrition. 2005;28:421–30.

[20] Grant CA, Clarke JM, Duguid S, Chaney RL. Selection and breeding of plant cultivars to minimize cadmium accumulation. Science of the Total Environment. 2008;390:301–310.

[21] Sugiyama M, Ae N, Arao T. Role of roots in differences in seed cadmium concentration among soybean cultivars-proof by grafting experiment. Plant and Soil. 2007b;295:1–11.

[22] Ishikawa S, Ae N, Sugiyama M, Murakami M, Arao T. Genotypic variation in shoot cadmium concentration in rice and soybean in soils with different levels of cadmium contamination. Soil Science & Plant Nutrition 2005;51:101–108.

[23] Arao T, Ishikawa S. Genotypic differences in cadmium concentration and distribution of soybeans and rice. JARQ Japan Agricultural Research Quarterly. 2006;40:21–30.

[24] Boggess SF, Willavize S, Koeppe DE. Differential response of soybean varieties to soil cadmium. Agronomy Journal. 1978;70:756–60.

[25] Bell MJ, McLaughlin MJ, Wright GC, Cruickshank A. Inter- and intra-specific variation in accumulation of cadmium by peanut, soybean and navy bean. Australian Journal of Agricultural and Resource Economics.1997;48:1151–1160.

[26] Wang P, Deng X, Huang Y, Fang X, Zhang J, Wan H, Yang C. Root morphological responses of five soybean [*Glycine max* (L.) Merr] cultivars to cadmium stress at young seedlings. Environmental Science and Pollution Research International. 2016 23(2): 1860-1872

[27] Benitez ER, Hajika M, Yamada T, Takahashi K, Oki N, Yamada N, Nakamura T, Kanamaru K. A major QTL controlling seed cadmium accumulation in soybean. Crop Science. 2010;50:1728–1734.

[28] Salazar MJ, Rodriguez JH, Nieto GL, Pignata ML. Effects of heavy metal concentrations (Cd, Zn and Pb) in agricultural soils near different emission sources on quality, accumulation and food safety in soybean [*Glycine max* (L.) Merrill]. Journal of Hazardous Materials. 2012;234:244–253.

[29] Zhuang P, Li ZA, Zou B, Xia HP, Wang G. Heavy metal contamination in soil and soybean near the Dabaoshan mine, South China. Pedosphere. 2013;23:298–304.

[30] Moon KH, Kim JY, Chang MI, Kim US, Kim SJ, Baek SH. Heavy metal and amino acid contents of soybean by application of sewage and industrial sludge. Journal of the Korean Society of Food Science and Nutrition. 2013;42:268–277.

[31] Li X, Zhang L, Li Y, Ma L, Bu N, Ma C. Changes in photosynthesis, antioxidant enzymes and lipid peroxidation in soybean seedlings exposed to UV-B radiation and/ or Cd. Plant Soil. 2012;352:377–387.

[32] Sugiyama M, Ae N, Arao T. Accumulation of large amounts of Cd in the root may limit the accumulation of Cd in edible above-ground portions of the plant. Plant and Soil. 2007a;295:1–11.

[33] Maejima Y, Makino T, Takano H, Kamiya T. Remediation of cadmium-contaminated paddy soils by washing with chemicals: Effect of soil washing on cadmium uptake by soybean. Chemosphere. 2007;67:748–754.

[34] Konotop Y, Mezsaros P, Matušíková I, Batsmanova L, Taran N. Application of nitrogen nutrition for improving tolerance of soybean seedlings to cadmium. Environmental and Experimental Biology. 2012;10:139–144.

[35] Rizwan AS, Bashir K, Hayat R, Akhtar MS. Effect of cadmium on soybean (*Glycine max* L) growth and nitrogen fixation. African Journal of Biotechnology. 2012;11:1886–1891.

[36] Chaney RL, Ryan JA, Li Y-M, Brown SL. Soil cadmium as a threat to human health. In: McLaughlin ML, Singh BR, editors. Cadmium in soils and plants. Dordrecht: Kluwer Academic Publishers; 1999. p. 219–256.

[37] Oliveira JA, Oliva MA, Cambraia J, Venegas VHA. Absorption, accumulation and distribution of cadmium by two soybean cvs. Revista Brasileira de Fisiologia Vegetal. 1994;6:91–95.

[38] Kawasaki A, Oda H. Application of ^{113}Cd as a tracer in evaluation of cadmium uptake by soybeans under field conditions. Japanese Journal of Soil Science and Plant Nutrition. 2005;76:261–267.

[39] Kawasaki A, Oda H, Yamada M. Application of ^{113}Cd as a tracer in evaluation of cadmium uptake by soybeans under pot conditions. Japanese Journal of Soil Science and Plant Nutrition. 2004;75:667–672.

[40] Oda H, Senda S, Kawasaki A. Uptake and transport of Cd supplied at different growth stages in hydroponically cultured soybean plants. Biomedical Research on Trace Elements. 2004;15:289–291.

[41] Kobori G, Okazaki M, Motobayashi T, Matsukawa K. Differences in cadmium uptake and accumulation in three soybean cultivars. Pedologist. 2011;54:214–221.

[42] Pinto AP, Mota AM, de Varennes A, Pinto FC. Influence of organic matter on the uptake of cadmium, zinc, copper and iron by sorghum plants. Science of Total Environment. 2004;326:239–247.

[43] Zhang L, Wong MH. Environmental mercury contamination in China: Sources and impacts. Environmental International. 2007;33:108–121.

[44] Yadav SK. Heavy metals toxicity in plants: An overview on the role of glutathione and phytochelatins in heavy metal stress tolerance of plants. South African Journal of Botany. 2010;76:167–179.

[45] Lin YF, Aarts MG. The molecular mechanism of zinc and cadmium stress response in plants. Cellular and Molecular Life Science. 2012;69:3187–3206.

[46] Hall JL. Cellular mechanisms for heavy metal detoxification and tolerance. Journal of Experimental Botany. 2002;53:1–11.

[47] Thomine S, Wang R, Ward JM, Crawford NM, Schroeder JI. Cadmium and iron transport by members of a plant metal transporter family in *Arabidopsis* with homology to Nramp genes. Proceedings of the National Academy of Sciences USA. 2000;97:4991–4996.

[48] Clemens S, Palmgren MG, Kramer U. A long way ahead: Understanding and engineering plant metal accumulation. Trends in Plant Science. 2002;7:309–315.

[49] Salt DE, Prince RC, Pickering IJ, Raskin I. Mechanism of cadmium mobility and accumulation in Indian mustard. Plant Physiology. 1995;109:1427–1433.

[50] Reddy MR, Dunn SJ. Heavy metal absorption by soybean on sewage sludge treated soil. Journal of Agricultural and Food Chemistry. 1986;34:750–753.

[51] Arao T, Ae N. Screening of genotypes with low cadmium content in soybean seed and rice grains. Plant nutrition: food security and sustainability of agro-ecosystems through basic and applied research. Fourteenth International Plant Nutrition Colloquium. Germany: Hannover; 2001. p. 292–293.

[52] Morrison M. Cadmium concentration in Canadian soybean. The Pulse Beat. Manitoba Pulse Growers Association. 2005; 47: p. 33–34.

[53] Jegadeesan S, Yu KF, Poysa V, Gawalko E, Morrison MJ, Shi C, Cober E. Mapping and validation of simple sequence repeat markers linked to a major gene controlling seed cadmium accumulation in soybean [*Glycinemax* (L.) Merr]. Theoretical and Applied Genetics. 2010;121:283–294.

[54] Guo Y, Marschner H. Genotypic differences in uptake and translocation of cadmium in bean and maize inbred lines. Z Pflanzenern Bodenk. 1996;159:55–60.

[55] Grant CA, Bailey LD, McLaughlin MJ, Singh BR. Management factors which influence cadmium concentration in crops. In: McLaughlin MJ, Singh BR, editors. Cadmium in soils and plants. Dordrecht: Kluwer Academic Publishers; 1999. p. 151–198.

[56] Archambault DJ, Marentes E, Buckley W, Clarke J, Taylor GJ. A rapid seedling-based bioassay for identifying low cadmium accumulating individuals of durum wheat (*Triticum turgidum* L.). Euphytica. 2001;117:175–182.

[57] Stolt P, Asp H, Hultin S. Genetic variation in wheat cadmium accumulation on soils with different cadmium concentrations. Journal of Agronomy and Crop Science. 2006;192:201–208.

[58] Cregan PB, Jarvick T, Bush AL, Shoemaker RC, Lark KG, Kahler AL, Kaya N, Van Toai TT, Lohnes DG, Chung J, Specht JE. An integrated genetic map of the soybean genome. Crop Science. 1999;39:1464–1490.

[59] Peter S, Nirit B, Ĺubomír R, Patrik M, Terézia G, Nadine S, Jana L, Jana M, Ildikó M. Cd accumulation potential as a marker for heavy metal tolerance in soybean, Israel Journal of Plant Sciences. 2015. DOI: 10.1080/07929978.2015.1042307.

[60] Vollmann J, Losˇa´k T, Pachner M, Watanabe D, Musilova´ L, Hlusˇek J. Soybean cadmium concentration: validation of a QTL affecting seed cadmium accumulation for improved food safety. Euphytica. 2015;203:177–184.

[61] Wang Y, Yu KF, Poysa V, Shi C, Zhou YH. A single point mutation in GmHMA3 affects cadmium (Cd)translocation and accumulation in soybean seeds. Molecular Plant. 2012a;5:1154–1156.

[62] Schmutz J, et al. Genome sequence of the palaeopolyploid soybean. Nature. 2010;463:178–183.

[63] Morel M, Crouzet J, Gravot A, Auroy P, Leonhardt N, Vavasseur A, Richaud P. AtHMA3, a P1B-ATPase allowing Cd/Zn/Co/Pb vacuolar storage in *Arabidopsis*. Plant Physiology. 2009;149:894–904.

[64] Ueno D, Yamaji N, Kono I, Huang CF, Ando T, Yano M, Ma JF. Gene limiting cadmium accumulation in rice. Proceedings of the National Academy of Sciences USA. 2010;107:16500–16505.

[65] Xiang C, Oliver DJ. Glutathione metabolic genes coordinately respond to heavy metals and jasmonic acid in *Arabidopsis*. Plant Cell. 1998;10:1539–1550.

[66] Suzuki N, Koizumi N, Sano H. Screening of cadmium responsive genes in *Arabidopsis thaliana* reveals protein denaturation and oxidative stresses to be critical components of cadmium toxicity. Plant, Cell & Environment. 2001;24:1177–1188.

[67] Louie M, Kondor N, Dewitt JG. Gene expression in cadmium tolerant *Datura inoxia*: Detection and characterization of cDNAs induced in response to $Cd2^+$. Plant Molecular Biology. 2003;52:81–89.

[68] Claudia J, Hirofumi N, Heribert H. Heavy metal stress. Activation of distinct mitogen-activated protein kinase pathways by copper and cadmium. Plant Physiology. 2004;136:3276–3283.

[69] Cheng-Ri Z, Takashi I, Yoshiharu S, Yuriko K, Yuji S, Takashi H, Shigeru S, Nozomu S, Daisuke S, Hiroyuki K. Comparative transcriptomic characterization of aluminium,

sodium chloride, cadmium and copper rhizotoxicities in *Arabidopsis thaliana*. BMC Plant Biology. 2009;9:32. Doi:10.1186/1471-2229-9-32.

[70] Cataldo DA, Garland TR, Wildung RE. Cadmium uptake kinetics in intact soybean plants. Plant Physiology. 1983;73:844–848.

[71] Astolfi S, Zuchi S, Chiani A, Passera C. In vivo and in vitro effects of cadmium on H $^{+}$ATPase activity of plasma membrane vesicle from oat (*Avena sativa* L.) roots. Journal of Plant Physiology. 2003;160:387–393.

[72] Serrano R. Structure and function of plasma membrane ATPase. Plant Molecular Biology. 1989;40:61–94.

[73] Williams LE, Pittman JK, Hall JL. Emerging mechanisms for heavy metal transport in plants. Biochimica et Biophysica Acta. 2000;1465:104–126.

[74] Benitez ER, Hajika M, Takahshi R. Single-base substitution in P_{1B}-ATPase. Gene is associated with a major QTL for seed cadmium concentration in soybean. Journal of Heredity. 2012;103:278–286.

[75] Matsumura H, Kitajima H, Akada S, Abe J, Minaka N, Takahashi R. Molecular cloning and linkage mapping of cryptochrome multigene family in soybean. Plant Genome. 2009;2:271–281.

[76] Mills RF, Krijger GC, Baccarini PJ, Hall JL, Williams LE. Functional expression of AtHMA4, a P1B-type ATPase of the Zn/Co/Cd/Pb subclass. The Plant Journal. 2003;35:164–176.

[77] Gravot A, Lieutaud A, Verret F, Auroy P, Vavasseur A, Richaud P. AtHMA3, a plant P1B-ATPase, functions as a Cd/Pb transporter in yeast. FEBS Letters. 2004;561:22–28.

[78] Wang Y, Xiao X, Zhang T, Kang H, Zeng J, Fan X, Sha L, Zhang H, Yu K, Zhou Y. Cadmium treatment alters the expression of five genes at the *Cda1* locus in two Soybean cultivars [*Glycine max* (L.) Merr]. The Scientific World Journal. 2014. Article ID 979750, doi:10.1155/2014/979750.

[79] Wang Y, Yu K, Poysa V, Shi C, Zhou Y. Selection of reference genes for normalization of qRT-PCR analysis of differentially expressed genes in soybean exposed to cadmium. Molecular Biology Reports. 2012b;39:1585–1594.

[80] Apel K, Hirt H. Reactive oxygen species: metabolism, oxidative stress, and signal transduction. Annual Review of Plant Biology. 2004;55:373–399.

[81] Mendoza-Soto AB, Sanchez F, Hernandez G. MicroRNAs as regulators in plant metal toxicity response. Frontiers in Plant Science. 2012;3:105.

[82] Bartel DP. MicroRNAs: genomics, biogenesis, mechanism, and function. Cell. 2004;116:281–297.

[83] Brodersen P, Sakvarelidze-Achard L, Bruun-Rasmussen M, Dunoyer P, Yamamoto YY, et al. Widespread translational inhibition by plant miRNAs and siRNAs. Science. 2008;320:1185–1190.

[84] Bartel DP. MicroRNAs: target recognition and regulatory functions. Cell. 2009;136:215–233.

[85] Mallory AC, Vaucheret H. Functions of microRNAs and related small RNAs in plants. Nature Genetics. 2006;38Suppl:S31–36.

[86] Ruiz-Ferrer V, Voinnet O. Roles of plant small RNAs in biotic stress responses. Annual Review of Plant Biology. 2009;60:485–510.

[87] Sunkar R, Zhu JK. Novel and stress-regulated microRNAs and other small RNAs from *Arabidopsis*. Plant Cell. 2004;16:2001–2019.

[88] Sunkar R, Kapoor A, Zhu JK. Posttranscriptional induction of two Cu/Zn superoxide dismutase genes in *Arabidopsis* is mediated by downregulation ofmiR398 and important for oxidative stress tolerance. Plant Cell. 2006;18:2051–2065.

[89] Ding YF, Zhu C. The role of microRNAs in copper and cadmium homeostasis. Biochemical and Biophysical Research Communications. 2009;386:6–10.

[90] Xie FL, Huang SQ, Guo K, Xiang AL, Zhu YY, et al. Computational identification of novel microRNAs and targets in *Brassica napus*. FEBS Letters. 2007;581:1464–1474.

[91] Huang SQ, Xiang AL, Che LL, Chen S, Li H, et al. A set of miRNAs from *Brassica napus* in response to sulphate deficiency and cadmium stress. Plant Biotechnology Journal. 2010;8:887–899.

[92] Huang SQ, Peng J, Qiu CX, Yang ZM (2009). Heavy metal-regulated new microRNAs from rice. Journal of Inorganic Biochemistry. 2009;103:282–287.

[93] Zhou ZS, Song JB, Yang ZM. Genome-wide identification of *Brassica napus* microRNAs and their targets in response to cadmium. Journal of Experimental Botany. 2012;63:4597–4613.

[94] Fang X, Zhao Y, Ma Q, Huang Y, Wang P, Zhang J, Nian H, Yang C. Identification and comparative analysis of a cadmium tolerance-associated miRNAs and their targets in two soybean genotypes. PLoS ONE. 2013;8:e81471. doi:10.1371/journal.pone.0081471.

[95] Franke R, Hemm MR, Denault JW, Ruegger MO, Humphreys JM, et al. Changes in secondary metabolism and deposition of an unusual lignin in the ref8mutant of *Arabidopsis*. The Plant Journal. 2002;30:47–59.

[96] Marmiroli M, Antonioli G, Maestri E, Marmiroli N. Evidence of the involvement of plant ligno-cellulosic structure in the sequestration of Pb: An X-ray spectroscopy-based analysis. Environmental Pollution. 2005;134:217–227.

[97] Elobeid M, Gobel C, Feussner I, Polle A. Cadmium interferes with auxin physiology and lignification in poplar. Journal of Experimental Botany. 2012;63:1413–1421.

[98] Finger-Teixeira A, Ferrarese Mde L, Soares AR, da Silva D, Ferrarese-Filho O. Cadmium-induced lignification restricts soybean root growth. Ecotoxicology and Environmental Safety. 2010;73:1959–1964.

[99] Yang Y-J, Cheng L-M, Liu Z-H. Rapid effect of cadmium on lignin biosynthesis in soybean roots. Plant Science. 2007;172:632–639.

[100] Akiyoshi DE, Klee H, Amasino RM, Nester EW, Gordon MP. T-DNA of *Agrobacterium tumefaciens* encodes an enzyme of cytokinin biosynthesis. Proceedings of the National Academy of Sciences USA. 1984;81:5994–5998.

[101] Barry GF, Rogers SG, Fraley RT, Brand L. Identification of a cloned cytokinin biosynthetic gene. Proceedings of the National Academy of Sciences USA. 1984;81:4776–4780.

[102] Merewitz EB, Gianfagna T, Huang B. Photosynthesis, water use, and root viability under water stress as affected by expression of SAG12-ipt controlling cytokinin synthesis in *Agrostis stolonifera*. Journal of Experimental Botany. 2011;62:383–395.

[103] Kuderova A, Urbankova I, Valkova M, Malbeck J, Brzobohaty B, et al. Effects of conditional IPT-dependent cytokinin overproduction on root architecture of *Arabidopsis* seedlings. Plant and Cell Physiology. 2008;49:570–582.

Chapter 5

Antioxidants Properties and Effect of Processing Methods on Bioactive Compounds of Legumes

Ma. del Socorro López-Cortez,
Patricia Rosales-Martínez,
Sofía Arellano-Cárdenas and
Maribel Cornejo-Mazón

Additional information is available at the end of the chapter

http://dx.doi.org/10.5772/63757

Abstract

Extensive research has proven that fruits and vegetables contribute significantly to the body supply of bioactive compounds due to their antioxidant activity to protect organisms against harmful effects of oxygen radicals. A special case is the legumes that are also rich source of proteins, dietary fiber, micronutrients, and bioactive phytochemicals. Many legume species are still an irreplaceable source of dietary proteins for humans, especially in the mainly vegetarian diets of developing countries. Incorporation of leguminous seeds into the human diet can offer protective effects against chronic diseases because they contain a number of bioactive substances including phenolics that can increase protein digestibility and mineral bioavailability. However, technological processing and seed germination can impact the levels of natural endogenous antioxidants (e.g., phenolics, tocopherols; vitamin C) in leguminous seeds. Therefore, this chapter is a review about reports of antioxidant properties and their relationship with their total phenolic content of the most commonly consumed legumes. Researches about changes in the content of natural antioxidants during technological processing are included as well as some clinical reports concerning to the health benefits offered by legumes of higher consumption.

Keywords: legumes, beans, soybeans, phenolics compounds, antioxidants

1. Introduction

Food legume crops are considered vital for agriculture in developing countries for their nutritional value in which rely both the producers and consumers. Food legumes are an important source of protein and minerals, complementing the diet when combined with cereals. In agronomical terms, legumes crops serve as rotation crops with cereals, increasing the amount of nitrogen and also reducing soil pathogens [1]. Added to this, extensive research has proven that fruits and vegetables contribute significantly to the body supply of bioactive compounds due to their antioxidant activity attributed to the phenolic compounds that are known to protect organisms against harmful effects of oxygen radicals. A special case is legumes that are also rich source of proteins, dietary fiber, micronutrients, and bioactive phytochemicals. Experimental, epidemiological, and clinical studies show correlations between the consumption of food legumes and decreasing incidence of several diseases, such as cancer, cardiovascular diseases, obesity, and diabetes [2–4]. The antioxidant capacity [5] and the antimutagenic [6, 7], apoptosis-related [8], and antiproliferative effects of legumes are associated with the presence of phenolic compounds [9, 10].

The antioxidant capacity of legumes is within a wide range, because it depends on the biological variety of the plant and its origin. On the other hand, technological processing and seed germination can impact the levels of natural endogenous antioxidants (e.g., phenolics, tocopherols; vitamin C) in leguminous seeds. However, food processing not only improves flavor and palatability of foods but also increases the bioavailability of nutrients, by inactivating antinutritional factors, growth inhibitors, and hemagglutinins [11]. Legumes must be cooked—typically by boiling process—before consumption, because it changes the chemical composition and physical characteristics, such as flavor, color, and biological active components. To accelerate the cooking process, legumes should be soaked prior to boiling. Other cooking alternatives include pressure boiling and steaming. Moreover, a high-quality product might be obtained using high-pressure cooking technology [12].

This chapter is a review about some relevant reports about the antioxidant properties and their relationship with their total phenolic content of the most commonly consumed legumes, as well as some researches about changes in the content of natural antioxidants during technological processing and some clinical reports concerning to the health benefits offered by legumes of higher consumption.

2. Legumes description

Legumes belong to the family Leguminosae, one of the most important families in Dicotyledons, including around 700 genera and 20,000 species [13]. Leguminosae or Fabaceae is the third most populous family of flowering plants (behind Asteraceae and Orchidaceae) and include important pasture, grain, and agro-forestry species [14].

Legume is a plant characterized by edible seeds, borne in pods that often open along two seams, by pea-shaped flowers, and by compound stipulate leaves [15]. These include alfalfa, clover, lupins, green beans and peas, peanuts, soybeans, dry beans, broad beans, dry peas, chickpeas, and lentils, and those of them represent an important component of the human diet in several areas of the world, where they complement the lack of proteins from cereals, roots, and tubers [16]. Legumes, such as lentils, chickpeas, and beans, have been cultivated for millennia all around the world; therefore, they have played a big role in many traditional cuisines of Asia, Central and South America, Middle East, and the Mediterranean, along with cereals (e.g., maize, barley, wheat, and rice) [17]. Although legume research is mostly dedicated to dry seeds [18], legumes are also consumed in salads as green vegetables (i.e., fresh pods, leaves, and seedlings); contain natural antioxidants; and are generally recognized as safe (GRAS) for human consumption. Proteins contained in legumes can counteract the oxidative effects of free radicals in biomolecules (e.g., DNA, lipids, and other proteins). In general, legumes are considered to be a better source of nutrients than cereals, because of their low glycemic indexes and fat (2–5%) and high amount of proteins, fibers, and carbohydrates (55–60%), which might be the reason why legumes were considered beneficial in traditional medicine [19, 20].

2.1. Antioxidant properties of higher consumption legumes

Antioxidants in legumes, such as flavonoids, phenolic acids, lignans, and tannins, are abundant in the seed coats [21, 22]. These phenolic compounds have a number of favorable physiological properties that are beneficial against chronic diseases. Antioxidants are naturally present in leguminous seeds; however, technological processing and seed germination can diminish their presence [11]. Antioxidant activities and phenolic compounds in raw legumes have been reported in several earlier communications. The following section describes some relevant studies on the antioxidant properties of the most common legumes (common beans, soybean, lima bean, lentils, peanut, peas, and chickpea).

2.1.1. Common bean (Phaseolus vulgaris L.)

Common bean (*P. vulgaris* L.), a member of the Leguminosae family, is a grain consumed in considerable quantities around the world. It is a plant native to America, specifically to the Andean and Mesoamerican regions. Common beans are a good source of protein (16–33%), some vitamins, minerals, and complex carbohydrates [23]. They also contain secondary metabolites such as tannins, anthocyanins, phenolic compounds, and fiber. There is evidence that these compounds, identified like to phytochemicals, play an important role in prevention and treatment of certain diseases. For example, the lower incidence of colon cancer registered in Latin-American countries as compared with other countries is partially due to the higher consumption of common bean [24].

Some of the main phenolic compounds found in various types of beans and their physiological properties are the following [25–29]: flavonoids (i.e., caffeic, *p*-coumaric, ferulic, and sinapic esters) present in a methanolic black-bean seed coat extract are thought to diminish liver injury in animal models, as well as colon, breast, and prostate cancer proliferation. Polyphenols present in a hot water pinto-bean hull extract increase bone metabolism in mice. Tannins, also present in black beans, hamper cancer cell proliferation. Anthocyanins are found in black beans (i.e., delphinidin, petunidin, and malvidin), pinto beans (i.e., kaempferol), and pink beans (i.e., quercetin and kaempferol), although their physiological effects have not been reported.

Moreover, peptides released after enzymatic hydrolysis also act as antioxidants because the phenolic, indole, and imidazole groups contained in their amino acids function as proton donors that stabilize free radicals [30]. Particularly, total hydrolysates (TH) or peptides derived after enzymatic hydrolysis and fractioning procedures from protein leguminous such as chickpea, soybean, pea, lentil, mung bean, and common beans have demonstrated an important antioxidant and angiotensin-I converting enzyme (ACE) [31]. The ACE is a key element in the rennin angiotensin system (RAS) responsible for the control of blood pressure. Recently, [32] reported that the protein hydrolysates and peptidic fractions obtained from different varieties of common beans (*P. vulgaris*) have several biological activities, such as the antioxidant, antimicrobial (inhibit the growth of *Shigella dysenteriae*), and antihypertensive activities (*in vitro* and *in vivo*).

A comparative study of protein profile and potential bioactive peptides of improved common bean cultivars grown in Mexico and Brazil was carried out, and the major identified proteins were phaseolin, lectin, and protease and α-amylase inhibitors, and abundant peptides were identified by HPLCMS/MS with molecular masses ranging from 300 to 1500 Da [33]. Peptides from common bean proteins presented potential biological activities related to control of hypertension and type-2 diabetes. As inflammatory reactions often include the formation of tissue-damaging oxidation products, compounds with high antioxidant activity may inhibit inflammation. Results by Oomah et al. [34] with bean hulls support previous studies in which antioxidant and anti-inflammatory activities of extracts are associated with polyphenols capable of inhibiting COX and LOX [34–36]. Animal models of cellular activity also provided evidence for chemopreventive effects of black bean hull extracts.

There are also some studies relating the beneficial effect of whole bean and reduction of chronic diseases related to inflammation such as colon cancer [37] and diabetes mellitus [38]. The antioxidant capacity of protein hydrolysates and the effects on the markers of inflammation in lipopolysaccharide (LPS-induced RAW 264.7) macrophages were evaluated in common bean (*P. vulgaris* L.) varieties, Negro 8025 and Pinto Durango [23]. They concluded that hydrolysates from the common bean could be used to combat inflammatory and oxidative-associated diseases.

Furthermore, the influence of thermal processing (canning and open pot) of common beans (*P. vulgaris* L.) varieties Black 8025 (N), Bayo Victoria (BV), Pinto Durango (PD), and Pinto Saltillo (PS) in their chemical composition, and their antioxidant and anti-inflammatory activities in a human intestinal cell model, was evaluated [39]. They concluded that the effect of cooking on bioactive compounds from common beans is cultivar dependent, being more quantitative than qualitative as a consequence of the release of bonded phenolics. Although the thermal processing is partially degrading some phenolics, at the same time it is releasing other bonded polyphenols. Cooked beans have shown good antioxidant properties as the raw materials, and in some cases, even better than the raw beans.

2.1.2. *Soybean (Glycine max (L.) Merr. Fabaceae)*

Soybeans are one of the most produced commodities worldwide and are among the most important crops for human and animal consumption; however, only four countries (USA, Brazil, Argentina, and China) are responsible for providing nearly 90% of soybean seeds worldwide [40]. Soybean seeds [*Glycine max* (L.) Merr. *Fabaceae*] contain a significant amount of protein (~40%) and oil (~20%). The antioxidants of soybeans are represented by isoflavones, tocopherols, ascorbic acid, and some other compounds [41, 42].

When soybeans are processed into different foods, the particular antioxidants content of the produced foods may change depending on the processing procedure [43] and storage conditions [44] and differ from the initial antioxidant content in the soybeans [45].

This species is a widely used crop because of its valuable beneficial health effects on several chronic diseases, including the prevention of cancer (including breast, colon, and prostate cancers), osteoporosis, cardiovascular disease, and multiple conditions ameliorated by antioxidants [46–48]. On the other hand, for a practical application in the food industry, antioxidants should be first extracted. The efficiency of the extraction process affects the antioxidant capacity of the extract [49]. Studies on the extraction of the antioxidant activity in unfermented soybeans and vine have reported a variation of the total phenolic concentration when different solvents were used, which is due to differences in their polarities [50]. Limited information is available regarding the extraction of antioxidant compounds in fermented soybeans. However, significant higher concentration of phenolics was obtained after fermentation when compared to unfermented soybeans [51, 52]. Until now, a regular extraction protocol has not been established because of the complex nature of the soybeans and their wide range of antioxidants. Also, other factors such as the temperature and the nature of solvent might react unpredictably and alter the extraction efficiency [53].

It is well established that the insoluble-bound phenolic compounds have high antioxidant and antiradical activities when compared with those of soluble and free phenolic compounds [54]. Soybean has high contents of the soluble phenolics, such as isoflavone and anthocyanin, in comparison with other phytochemicals. It has been reported in several studies the phenolic compounds and antioxidant activity of soybean. Among soybeans containing various seed coat colors, black soybeans showed strong antioxidant properties using the α,α-diphenyl-β-picryl hydrazyl (DPPH), ferric-reducing antioxidant power (FRAP), and oxygen radical absorbance capacity (ORAC) methods [48]. Brown and black soybeans contain highly polymeric seed procyanidins and anthocyanins [55]. Furthermore, black soy beans were observed higher radical scavenging activities than those of green and yellow soy beans [56].

The measured distribution of antioxidant capacity in black soybeans depends on the assay technique: DPPH and FRAP methods show that the seed coat contributes to 90% of the entire antioxidant activity of the soybean; conversely, the ORAC method shows that the seed coat and dehulled part of the soybean contribute equally to the antioxidant capacity [12]. These results, although contradicting, are helpful for the creation of effective treatments.

The antioxidant activity and contents of various polyphenol classes in the seeds of 20 soybean hybrids were evaluated [57]. They found a positive linear correlation between the antioxidant

[18] Mitchell DC, Lawrence FR, Hartman TJ, Curran JM. Consumption of dry beans, peas, and lentils could improve diet quality in the US population. Journal of the American Dietetic Association. 2009; 109(5): 909–913. DOI: 10.1016/j.jada.2009.02.029.

[19] Guarrera PM. Traditional phytotherapy in Central Italy (Marche, Abruzzo, and Latium). Fitoterapia. 2005; 76(1): 1–25. DOI: 10.1016/j.fitote.2004.09.006.

[20] Rochfort S, Panozzo J. Phytochemicals for health, the role of pulses. Journal of Agricultural and Food Chemistry. 2007; 55(20): 7981–7994. DOI: 10.1021/jf071704w.

[21] Aberoumand A, Deokule S.S. Comparison of phenolic compounds of some edible plants of Iran and India. Pakistan Journal of Nutrition. 2008; 7(4): 582–585. DOI: 10.3923/pjn.2008.582.585.

[22] Amarowicz R, Pegg RB. Legumes as a source of natural antioxidants. European Journal of Lipid Science and Technology. 2008; 110(10): 865–878. DOI: 10.1002/ejlt.200800114.

[23] Oseguera-Toledo ME, Gonzalez de Mejia E, Dia VP, Amaya-Llano SL. Common bean (*Phaseolus vulgaris* L.) hydrolysates inhibit inflammation in LPS-induced macrophages through suppression of NF-κB pathways. Food Chemistry. 2011; 127(3): 1175–1185. DOI: 10.1016/j.foodchem.2011.01.121.

[24] Reynoso-Camacho R, Ríos Ugalde MdC, Torres Pacheco I, Acosta Gallegos JA, Palomino Salinas AC, Ramos Gómez M, González Jasso E, Guzmán Maldonado SH. Common vean (*Phaseolus vulgaris* L.) consumption and its effects on colon cancer in Sprague–Dawley rats. Agricultura Tecnica en Mexico. 2007; 33(1): 43–52. DOI: 0568-2517.

[25] Onyeneho SN, Hettiarachchy NS. Effect of navy bean hull on the oxidative stability of soy and sunflower oils. Journal of Agricultural and Food Chemistry. 1991; 39(10): 1701–1704. DOI: 10.1021/jf00010a600.

[26] Beninger CW, Hosfield GL. Antioxidant activity of extracts, condensed tannin fractions, and pure flavonoids from *Phaseolus vulgaris* L. seed coat color genotypes. Journal of Agricultural and Food Chemistry. 2003; 51(27): 7879–7883. DOI: 10.1021/jf0304324.

[27] Luthria DL, Pastor-Corrales MA. Phenolic acids content of fifteen dry edible bean (*Phaseolus vulgaris* L.) varieties. Journal of Food Composition and Analysis. 2006; 19(2–3): 205–211. DOI: 10.1016/j.jfca.2005.09.003.

[28] Ross KA, Beta T., Arntfield SD. A comparative study on the phenolic acids identified and quantified in dry beans using HPLC as affected by different extraction and hydrolysis methods. Food Chemistry. 2009; 113 (1): 336–344. DOI: 10.1016/j.foodchem.2008.07.064.

[29] Lin LZ, Harnly JM, Pastor-Corrales MS, Luthria DL. The polyphenolic profiles of common bean (*Phaseolus vulgaris* L.). Food Chemistry. 2008; 107(1): 399–410. DOI: 10.1016/j.foodchem.2007.08.038.

[30] Oomah BD, Corbé A, Balasubramanian P. Antioxidant and anti-inflammatory activities of bean (*Phaseolus vulgaris* L.) hulls. Journal of Agricultural and Food Chemistry. 2010; 58(14): 8225–8230. DOI: 10.1021/jf1011193.

[31] Han KH, Fukushima M, Ohba K, Shimada KI, Sekikawa M, Chiji H, Lee CH, Nakano M. Hepatoprotective effects of the water extract from adzuki bean hulls on acetaminophen-induced damage in rat liver. Journal of Nutritional Science and Vitaminology. 2004; 50(5): 380–383. DOI: 10.3177/jnsv.50.380.

[32] Cao JJ, Gregoire BR, Sheng X, Liuzzi JP. Pinto bean hull extract supplementation favorably affects markers of bone metabolism and bone structure in mice. Food Research International. 2010; 43(2): 560–566. DOI: 10.1016/j.foodres.2009.07.031.

[33] Bawadi HA, Bansode RR, Trappey A, Truax RE, Losso JN. Inhibition of Caco-2 colon, MCF-7 and Hs578T breast, and DU 145 prostatic cancer cell proliferation by water-soluble black bean condensed tannins. Cancer Letters. 2005; 218(2): 153–162. DOI: 10.1016/j.canlet.2004.06.021.

[34] Oomah BD, Corbé A, Balasubramanian P. Antioxidant and anti-inflammatory activities of bean (*Phaseolus vulgaris* L.) hulls. Journal of Agricultural and Food Chemistry. 2010; 58(14): 8225–8230. DOI: 10.1021/jf1011193.

[35] Han KH, Fukushima M, Ohba K, Shimada KI, Sekikawa M, Chiji H, Lee CH, Nakano M. Hepatoprotective effects of the water extract from adzuki bean hulls on acetaminophen-induced damage in rat liver. Journal of Nutritional Science and Vitaminology. 2004; 50(5): 380–383. DOI: 10.3177/jnsv.50.380.

[36] Cao JJ, Gregoire BR, Sheng X, Liuzzi JP. Pinto bean hull extract supplementation favorably affects markers of bone metabolism and bone structure in mice. Food Research International. 2010; 43(2): 560–566. DOI: 10.1016/j.foodres.2009.07.031.

[37] Boateng JA, Verghese M, Walker LT, Shackelford LA, Chawan CB. Inhibitory effects of selected dry beans (*Phaseolus* spp. L.) on azoxymethane-induced formation of aberrant crypt foci in Fisher 344 male rats. Nutrition Research. 2007; 27(10): 640–646. DOI: 10.1016/j.nutres.2007.07.004.

[38] Tormo MA, Gil-Exojo I, de Tejada AR, Campillo JE. White bean amylase inhibitor administered orally reduces glycaemia in type 2 diabetic rats. British Journal of Nutrition. 2006; 96(3): 539–544. DOI: 10.1079/BJN20061836.

[39] Moreno-Jiménez MR, Cervantes-Cardoza V, Gallegos-Infante JA, González-Laredo RF, Estrella I, García-Gasca TJ, Herrera-Carrera E, Díaz-Rivas JO, Rocha-Guzmán NE. Phenolic composition changes of processed common beans: their antioxidant and anti-

inflammatory effects in intestinal cancer cells. Food Research International. 2015; 76(1): 79–85. DOI: 10.1016/j.foodres.2014.12.003.

[40] Masuda T, Goldsmith PD. World soybean production: area harvested, yield, and long-term projections. International Food and Agribusiness Management Review. 2009; 12(4): 143–162. DOI: http://purl.umn.edu/92573.

[41] Malenčić D, Popović M, Miladinović J. Phenolic content and antioxidant properties of soybean (*Glycine max* (L.) Merr.) seeds. Molecules. 2007; 12(3): 576–581. DOI: 10.3390/12030576.

[42] Barnes S. The biochemistry, chemistry and physiology of the isoflavones in soybeans and their food products. Lymphatic Research and Biology. 2010; 8(1): 89–98. DOI: 10.1089/lrb.2009.0030.

[43] Xu B, Chang SK, Liu Z, Yuan S, Zou Y, Tan Y. Comparative studies on the chemical and cell-based antioxidant activities and antitumor cell proliferation properties of soy milk manufactured by conventional and commercial UHT methods. Journal of Agricultural and Food Chemistry. 2010; 8(6): 3558–3566. DOI: 10.1021/jf903796c.

[44] Rau De Almeida Callou K, Sadigov S, Lajolo FM, Genovese MI. Isoflavones and antioxidant capacity of commercial soy-based beverages: effect of storage. Journal of Agricultural and Food Chemistry. 2010; 58(7): 4284–4291. DOI: 10.1021/jf904130z.

[45] Anderson RL, Wolf WJ. Compositional changes in trypsin inhibitors, phytic acid, saponins and isoflavones related to soybean processing. Journal of Nutrition. 1995; 125(3): 581S–588S. DOI:61604.

[46] Tsukamoto Y, Ichise H, Kakuda H, Yamaguchi M. Intake of fermented soybean (natto) increases circulating vitamin K2 (menaquinone-7) and -carboxylatedosteocalcin concentration in normal individuals. Journal of Bone and Mineral Metabolism. 2000; 18(4): 216–222. DOI: 10.1007/s007740070023.

[47] Park D, Huang T, Frishman WH. Phytoestrogens as cardioprotective agents. Cardiology in Reviews. 2005; 13(1): 13–17. DOI: 10.1097/01.crd.0000126084.68791.32.

[48] Zhang RF, Zhang FX, Zhang MW, Wei ZC, Yang CY, Zhang Y, Tang XJ, Deng YY, Chi JW. Phenolic composition and antioxidant activity in seed coats of 60 Chinese black soybean (*Glycine max* L. Merr.) varieties. Journal of Agricultural and Food Chemistry. 2011; 59(11): 5935–5944. DOI: 10.1021/jf201593n.

[49] Hinneburg I, Neubert RHH. Influence of extraction parameters of the phytochemical characteristics of extracts from buckwheat (*Fagopyrum esculentum*) herb. Journal of Agricultural and Food Chemistry. 2005; 53(1): 3–7. DOI: 10.1021/jf049118f.

[50] Naczk M, Shahidi F. Phenolics in cereal, fruits and vegetables: occurrence, extraction and analysis. Journal of Pharmaceutical and Biomedical Analysis. 2006; 41(5): 1523–1542. DOI: 10.1016/j.jpba.2006.04.002.

[51] McCue P, Shetty K. Role of carbohydrate-claving enzyme in phenolic antioxidant mobilization from whole soybean fermented with *Rhizopus oligosporus*. Food Biotechnology. 2003; 17(1): 27–37. DOI: 10.1081/FBT-120019982.

[52] Wardhani DH, Vázquez JA, Pandiella SS. Mathematical modeling of the development of antioxidant activity in soybeans fermented with *Aspergillus oryzae* and *Aspergillus awamori* in the solid state. Journal of Agricultural and Food Chemistry. 2009; 57(2): 540–544. DOI: 10.1021/jf802492s.

[53] Liu FF, Ang CYW, Springer D. Optimization of extraction conditions for active components in *Hypericum perforatum* using response surface methodology. Journal of Agricultural and Food Chemistry. 2000; 48(8): 3364–3371. DOI: 10.1021/jf991086m.

[54] Madhujith J, Shahidi F. Antioxidant potential of barley as affected by alkaline hydrolysis and release of insoluble bound phenolics. Food Chemistry. 2009; 117(4): 615–620. DOI: 10.1016/j.foodchem.2009.04.055.

[55] Takahata Y, Ohnishi-Kameyama M, Furuta S, Takahashi M, Suda I. Highly polymerized procyanidins in brown soybean seed coat with a high radical-scavenging activity. Journal of Agricultural and Food Chemistry. 2001; 49(12): 5843–5847. DOI: 10.1021/jf010307x.

[56] Kumar V, Rani A, Dixit AK, Pratap D, Bhatnagar D. A comparative assessment of total phenolic content, ferric reducing-anti-oxidative power, free radical-scavenging activity, vitamin C and isoflavones content in soybean with varying seed coat colour. Food Research International. 2010; 43(1): 323–328. DOI: 10.1016/j.foodres.2009.10.019.

[57] Malenčić D, Maksimović Z, Popović M, Miladinović J. Polyphenol contents and antioxidant activity of soybean seed extracts. Bioresource Technology. 2008; 99(14): 6688–6691. DOI: 10.1016/j.biortech.2007.11.040.

[58] Salgado AG, Gepts P, Debouck DG. Evidence for two gene pools of the lima bean, *Phaseolus lunatus* L., in the Americas. Genetic Resources and Crop Evolution. 1995; 42(1): 15–28. DOI: 10.1007/s10709-015-9863-0.

[59] Serrano-Serrano ML, Hernández-Torres J, Castillo-Villamizar G, Debouck DG, Sánchez MIC. Gene pools in wild Lima bean (*Phaseolus lunatus* L.) from the Americas: evidences for an Andean origin and past migrations. Molecular Phylogenetics and Evolution. 2010; 54(1): 76–87. DOI: 10.1016/j.ympev.2009.08.028.

[60] Agostini-Costa TDS, Teodoro AFP, Alves RDBDN, Braga LR, Ribeiro IF, Silva JP, Quintana LG, Burle ML. Total phenolics, flavonoids, tannins and antioxidant activity of lima beans conserved in a Brazilian Genebank. Ciência Rural. 2015; 45(2): 335–341. DOI: 10.1590/0103-8478cr20140030.

[61] Ezueh MI. Cultivation and utilization of minor food legumes in Nigeria. Tropical Grain Legume Bulletin. 1977; 10: 28–32. DOI: 10.1300/J108v06n02_08.

palntarum strain 299 V. Food Microbiology. 1995; 12: 487–495. DOI: 10.1016/S0740-0020(95)80133-2.

[112] Brenes A, Brenes J. Technological treatment of Legume: influence on their nutritional value. 2015. Available in: http://www.ucv.ve/fileadmin/user_upload/facultad_agronomia/Procesamiento_de_Granos_de_Leguminosas.pdf. [Accessed: 2015-12-12].

[113] Gómez-Villalva PE. Transformation and improving the nutritional value of pea flour by adding phytase enzyme [Doctoral Thesis]. Department of Physiology. Granada University. 2005.

[114] Hefnawy TH. Effect of processing methods on nutritional composition and anti-nutritional factors in lentils (*Lens culinaris*). Annals of Agricultural Science. 2011; 56(2): 57–61. DOI: 10.1016/j.aoas.2011.07.001.

[115] Somers DA, Samac DA, Olhoft PM. Recent advances in legume transformation. Plant Physiology. 2003; 131(3): 892–899. DOI: 10.1104/pp.102.017681.

[116] Wang N, Lewis MJ, Brennan O, Westby A. Effect of processing methods on nutrients and anti-nutritional factors in cowpea. Food Chemistry. 1997; 58(l–2): 59–68. DOI: 10.1016/S0308-8146(96)00212-9.

[117] Kozlowska H, Honke J, Sadowska J, Frías J, Vidal-Valverde C. Natural fermentation of lentils: influence of time, concentration and temperature on the kinetics of hydrolysis of inositol phosphates. Journal of the Science of Food and Agriculture. 1996; 71(3): 367–375. DOI: 10.1002/(SICI) 1097-0010(199607).

Chapter 6

Intercropping Promotes the Ability of Legume and Cereal to Facilitate Phosphorus and Nitrogen Acquisition through Root-Induced Processes

M. Latati , S. Benlahrech , M. Lazali , Tellah Sihem ,
G. Kaci , R. Takouachet , N. Alkama , F.Z. Hamdani ,
E.A. Hafnaoui , B. Belarbi , G. Ounane and
S.M. Ounane

Additional information is available at the end of the chapter

http://dx.doi.org/10.5772/63438

Abstract

Intercropping of cereal and legume can improve the use of resources for crop growth compared to cropping system. An increase in soil phosphorus (P) and nitrogen (N) acquisition by root-induced biochemical changes of intercropped species has been reported as key processes of facilitation and complementarily between both intercropping legumes and cereals. Indeed, the functional facilitation prevails over interspecific competition under nutrients limiting for crop growth. Results showed that P availability significantly increased in the rhizosphere of both species, especially in intercropping under the P-deficient soil conditions. This increase was associated with high efficiency efficiency in use of rhizobial, plant growth and resource use efficiency as indicated by higher land equivalent ratio (LER) and N nutrition index. In addition, the rhizosphere P availability and nodule biomass were positively correlated ($r^2 = 0.71^{**}$, and $r^2 = 0.62^{**}$) in the intercropped common bean grown at P-deficient soil. The increased P availability presumably improved biomass and yield in intercropping, although it mainly enhanced intercropped maize grain yield. Exploiting belowground parameters in a legume-cereal intercropping is likely necessary to maximize rhizosphere-interspecific interactions as a strategy to improve the symbiotic rhizobial efficiency and microbial activities, as a result of root-induced pH and N availability changes under low P soils.

Keywords: intercropping, symbiosis, legumes, cereals, phosphorus, Algeria

1. Introduction

Nitrogen (N) and phosphorus (P) are often considered to be the most important limiting factors, after water deficit and salinity, for plant growth and yield production in natural agroecosystems [1]. In cropping system and under stress conditions, the input of P and N via mineral fertilizers has been practiced to improve yielding agroecosystems [2–4]. However, the availability of P fertilizers is increasingly limited by the depletion of P mineral reserves with the growing food needs [5, 24]. Another approach is to increase the soil P availability that is often limited by adsorption on surfaces of mineral phases and fixation to cations such as Ca^{2+}, Al^{3+} or Fe^{2+} [6, 7].

Adopting sustainable technologies to better exploit soil nutrients resources, such as P and N, has been an interesting research challenge. Thus, the management of agricultural practices, including intercropped cultivation of cereals with legumes, is so far considered as one of the main agriculture sustainable components [2, 3, 6]. Recent studies reported that legumes-cereals in intercropping as compared to monocropping systems introduced greater environmental sources use efficiency for either plant growth or yields due to interspecific complementary, facilitation and competition between intercropped species [8–10, 17].

Increased acquisition of N has been mostly demonstrated in cereal-legume intercrops, compared to sole crops, only a few recent studies have reported the P or N-P interaction effect [6, 10, 11, 24]. Indeed, most of the former studies on cereal-legume intercropping implicitly assume that the legume enhances P and N acquisition by the cereal because of legumes' ability to increase large amounts of P-mobilizing compounds that ultimately increase P availability [6, 7, 12].

Root-induced some biological and chemical changes that can help to alter the rhizosphere processes of both intercropped legumes and cereals through (i) proton release by roots of N_2-fixing legumes [13, 14]; (ii) alkalization can also increase rhizosphere P availability in noncalcareous soils [7, 15]; and (iii) CO_2 emissions from the soil surface, which are the result of the overall activity of soil microorganisms and root-nodule symbionts, may be involved in the control of P availability in an alkaline soil [4].

In this context, fallow-cereal-rotation is the common cropping system for cereals production in Algeria. Actually, replacing fallow by legume crops in such farming systems of Algeria has become a strategic necessity for food security in a context of rising prices of food products [2, 13]. However, northern Algeria soils are among the most alkaline and calcareous soil in the Mediterranean conditions with high pH (7.5–8.5) and are considered Mediterranean zones [10, 16]. The following revision of the literature focuses on advantage of intercropping legumes-cereals under Algerian agroecosystems conditions.

2. Plant growth and nodulation under legumes-cereals intercropping

Although consistent progress has been made in exploring the intercropping cereal-legume advantages for better growth and productivity, this cropping system needs to be more deeply

http://dx.doi.org/10.5772/63438

investigated in order to point out abiotic stress tolerance traits such as those associated with the low nutrients availability in the soil [10]. The increase in cereals biomass and grain yield in association with a legume has been demonstrated for maize when it was grown intercropped with cowpea [7, 14] and also durum wheat in intercropping with faba bean. Legumes-cereals dual intercropping, provide the P and increase its availability for cereals [6, 10]. Recent studies show that total shoot dry weight of mixed cereals and legumes (above-ground biomass) was significantly higher in intercropping than in sole crop (**Figure 1**) [10]. Recent studies reported a significant increase in above-ground biomass of intercropped faba bean during continuous maize-faba bean intercropping for 9–10 years [12]. Legumes, with their adaptability to different cropping patterns and their ability to fix N_2, may offer opportunities to sustain increased plant biomass for intercropped species [2, 6, 7]. Several studies have addressed the effect of intercropping in increasing nodule growth [2, 18].

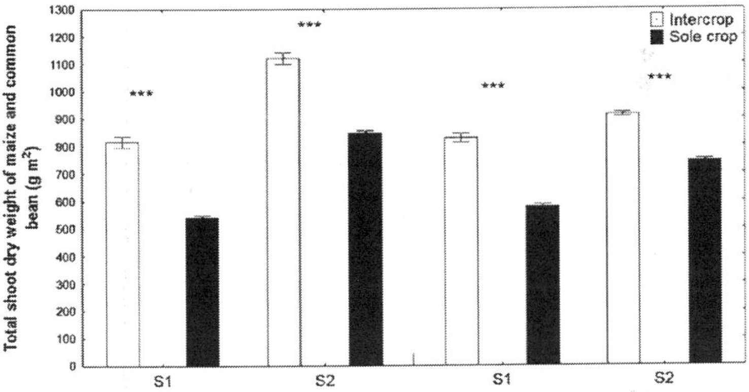

Figure 1. Total shoot dry weight (maize and common as intercrops or monocrops) per land area.

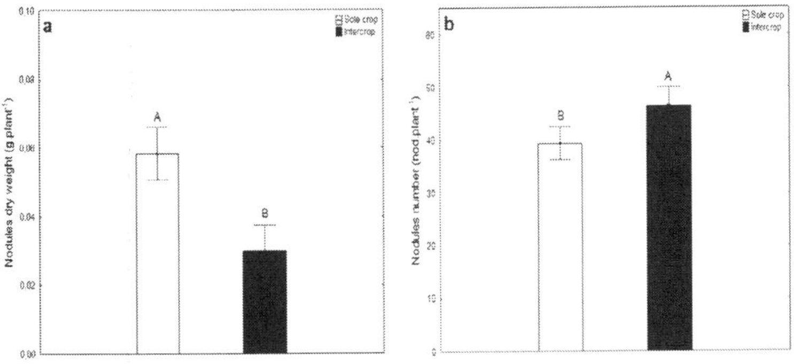

Figure 2. Dry weight of nodules (a) and number of nodules (b) for cowpea in sole cropping and intercropping.

However, nodule biomass for intercropped legumes were significantly decreased compared to monocropped legumes. A limited number of recent studies have addressed the same effect of intercropping on nodule growth for chickpea and common bean [2, 6]. In low P alkaline soil, it was reported a greater cowpea nodule number weight under intercropping with maize due to complementarily effect [7].

The decrease in nodule biomass was partly compensated by an increase in nodule number (**Figure 2a** and **b**) [7], which could be due to a change in the population of efficient rhizobial strains involved in root infection and efficient nodulation with higher nitrogenase activity [7, 19].

3. Increased efficiency in use of the rhizobial symbiosis (EURS)

The increase in the EURS of intercropped legumes in intercropping can be explained by interspecific competition for nitrogen use by the dual intercropping. Field research studies show a significant increase in N_2 fixation by common bean, as a result of competition with either durum wheat or with maize [2, 8].

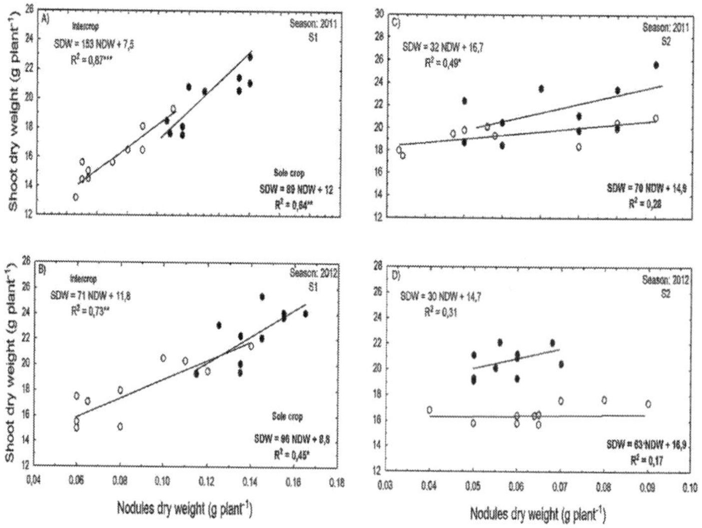

Figure 3. Efficiency in use of the rhizobial symbiosis in common bean as sole crops (filled circle) or intercrops (opened circle) under S1 (P deficient) and S2 (P sufficient) conditions.

An increase in EURS (mostly during low P availability: **Figure 3**) [10] indicating a tight relationship between legume N_2 fixation, growth and total grain yield. However, detecting differences in EURS between legumes grown in both sole and intercrops may offer an important clue in investigating key processes that influence P availability under P deficiency, where

http://dx.doi.org/10.5772/63438

legume's reliance to N_2 fixation presumably increased in parallel to a number of rhizosphere-induced changes (proton release, organic acids exudation, acid phosphatases, etc.) that contributed to increase P availability (**Figure 4**) [10] and growth [20].

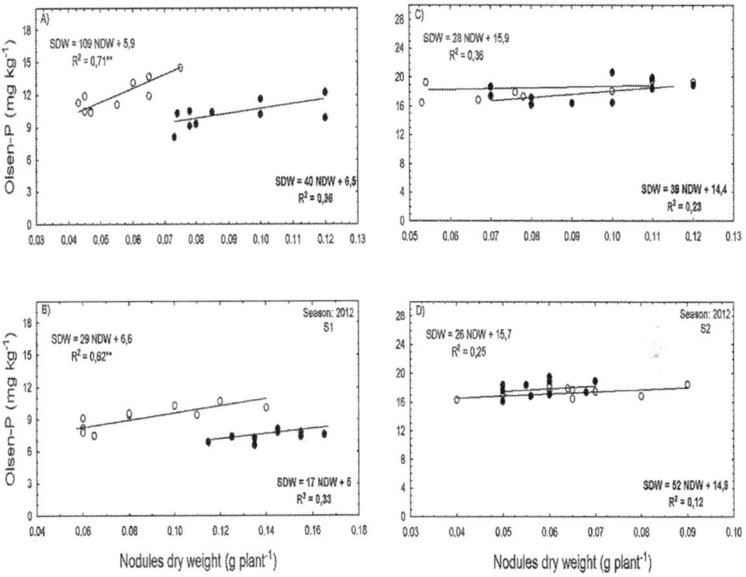

Figure 4. Correlation between rhizosphere soil Olsen P and nodule dry weight of common bean grown as sole crop (filled circle) or intercrop (opened circle) under S1 (P deficient) and S2 (P sufficient) conditions.

Recent studies were reported a high EURS of cowpea and common bean among intercrops treatment compared to corresponding EURS as sole crop, the increase in EURS by intercropping was significantly observed under low P conditions in either alkaline or calcareous soil [7, 10].

4. Phosphorus availability and root-induced changes

Several studies have reported the decline in the availability of P in the rhizosphere via root uptake during the crop cycle [11, 22]. Nevertheless, recent studies show an increase in P availability in the rhizosphere of intercropped legumes and cereals [22]. Recent researches reported an increase in inorganic P availability (Olsen-P) in the rhizosphere of both intercropped legumes and cereals [7, 10, 21]. Theses authors suggested that P deficiency can promote P availability through the root-induced processes (**Figure 5**) [7] in an alkaline soil, for example, rhizosphere acidification by legumes, nodules root respiration, exudation of phosphatases, carboxylates and/or indirectly through microbial activities [7, 14, 22].

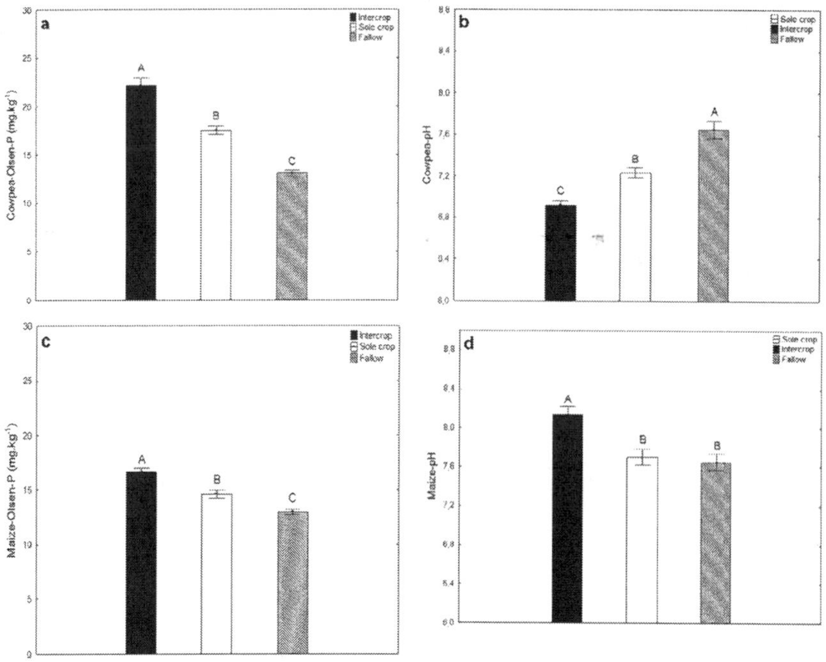

Figure 5. Olsen phosphorus (a, c) and pH (b, d) in the rhizosphere of cowpea and maize as sole crop and intercrop and in the fallow.

Indeed, nutrient limitation is the norm in native soils, especially in alkaline or calcareous soils, including many aridisols and some entisols, which are characterized by poor availability of P and N are less favorable than in most managed systems [2, 3].

Recent studies observed, under field experiments, a significant increase in P availability in the rhizosphere of both common bean and cowpea intercropped with maize [7, 10]. An increase in P availability was reported to (i) an acidification in the rhizosphere of cowpea and common bean in intercropping, (ii) alkalization in the rhizosphere of maize, it was significant only for the maize in intercropping and (iii) an increase of nodules-root respiration in intercropping compared to the monocropping system. Few research studies suggest that the availability of P in the rhizosphere is affected not only by changes in pH, but also by interacting with other root-induced changes such as an increase in EURS and C-CO_2 flux from microbial and root activity (**Figure 6**) [7].

However, species interactions resulted in an increase in growth only for maize in the alkaline and calcareous low P soil. In the other hand, these authors were reported a significant correlation between nodule biomass and Olsen-P in the rhizosphere of intercropped common bean in low P conditions indicates a positive effect of nodule growth in altering rhizosphere P availability.

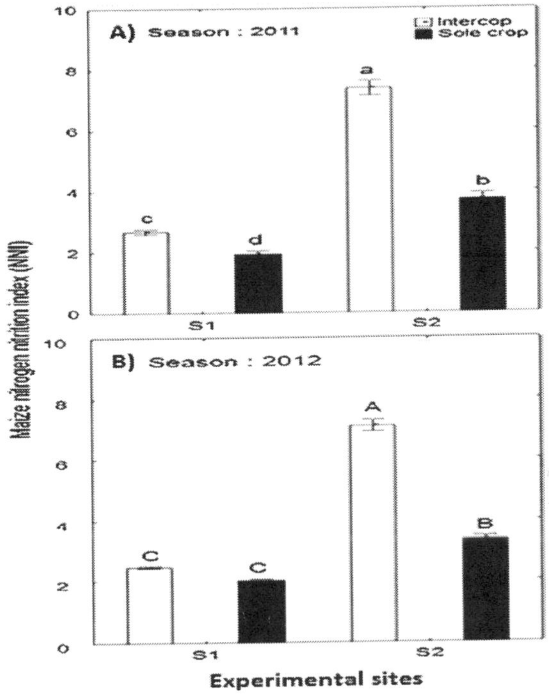

Figure 6. A, B: Nitrogen nutrition index (NNI) of maize as sole crop or intercrop under S1 (P-deficient) and S2 (P-sufficient) conditions.

5. Phosphorus and nitrogen nutrition under legumes-cereals intercropping

For intercropped cereals, an increase in P and N concentration and plant biomass, associated with an increase in grain yield, is assumed to result from the positive effect of legumes on P availability [2, 7]. Li et al. [23] and Latati et al. [10] reported an improvement in the growth of intercropped maize by improved P nutrition. For intercropped legumes, no facilitation was observed; these authors suggest that phosphatase activity produced by either chickpea or common bean increased the mineralization of organic P and its absorption by the associated maize.

In one of last research study, results showed that the EURS was significantly increased in both common bean and cowpea intercropped with maize. Such an increase is associated with high N and P availability in the rhizosphere of common bean and cowpea in intercropping, as a result, increase in N (**Table 1**) [2] and P uptake in shoot and seed of intercropped maize, especially for under low P and N conditions [7, 10].

Sites treatment	Crop treatment	Common bean Shoot N concentration (mg g^{-1})	Common bean Root N concentration (mg g^{-1})	Common bean Seed N concentration (mg g^{-1})	Maize Shoot N concentration (mg g^{-1})	Maize Root N concentration (mg g^{-1})	Maize Seed N concentration (mg g^{-1})
S1	Intercrop	45.3 ± 0.1 b	14.2 ± 0.3 c	54.7 ± 0.4 d	29.6 ± 0.2 a	4.1 ± 0.04 c	23.7 ± 0.2 b
S1	Sole crop	58.2 ± 0.08 a	17.5 ± 0.2 b	59.5 ± 1.1 b	25.8 ± 0.3 b	3.8 ± 0.1 c	18.4 ± 0.6 d
S2	Intercrop	38.4 ± 0.4 d	15.4 ± 0.1 c	71.3 ± 0.2 a	30 ± 0.2 a	5.7 ± 0.06 b	19.3 ± 0.08 d
S2	Sole crop	36.2 ± 0.3 e	11.3 ± 0.6 d	57.2 ± 0.08 c	25.3 ± 0.2 b	7.5 ± 0.08 a	20.8 ± 0.1 c
S3	Intercrop	40.7 ± 0.06 c	17.1 ± 0.1 b	58.9 ± 0.2 bc	23.8 ± 0.4 c	3.6 ± 0.1 cd	27.5 ± 0.08 a
S3	Sole crop	37.6 ± 0.1 d	21.4 ± 0.2 a	58.4 ± 0.2 bc	19.7 ± 0.3 d	3.2 ± 0.2 d	20.6 ± 0.1 c
		Shoot N concentration *p* Values	**Root N concentration *p* Values**	**Seed N concentration *p* Values**	**Shoot N concentration *p* Values**	**Root N concentration *p* Values**	**Seed N concentration *p* Values**
Sites		<0.001	<0.001	<0.001	<0.001	<0.001	<0.001
Crop		<0.001	<0.001	<0.001	<0.001	<0.001	<0.001
Sites × crop		<0.001	<0.001	<0.001	0.34	<0.001	<0.001

Table 1. Nitrogen concentration in shoots, roots and seed for maize and common bean in sole crop and intercropping.

Indeed, legume had a positive effect on interspecific competition through nitrogen partitioning with the intercropped cereal via increased of N_2 fixation under intercropping system. Analyzing the nitrogen nutrition index (NNI) in maize also added value in explaining the intercropping grain yield advantage and resource use improvement. This is clearly seen under P-deficient soil where intercropped maize (compared to sole-cropped maize) increased maize NNI nutrition (**Figure 6**) [10]. Enhancing the maize NNI appears to be in agreement with the increased total N uptake under P-deficient soil, but to a more extent in P-sufficient soil where higher maize root biomass would have greatly competed for soil N uptake [15].

6. Advantage of intercropping on grain yield and nutrients uptake

In terms of grain yield, intercropping had a positive and significant effect on the total grain yield as attested by the higher LER (yield advantage) over that found in sole cropping. This observation under field experiments indicates an increased crop performance and resource use efficiency of limiting resources (**Table 2**) [10], but to a larger extent in the P-deficient soil where LER of grain yield and total P and N were significantly higher compared to P-sufficient soil [10, 14].

Season	Exper. sites	LER (yield)	LER (TDW)	LER (N)	LER (P)
2011	S1	2.45 ± 0.2^{a}	2.02 ± 0.09^{ab}	2.06 ± 0.1^{b}	2.1 ± 0.1^{a}
	S2	1.67 ± 0.1^{b}	1.91 ± 0.1^{ab}	2.67 ± 0.07^{a}	1.76 ± 0.009^{ab}
2012	S1	2.71 ± 0.1^{a}	2.3 ± 0.1^{a}	2.02 ± 0.04^{b}	1.93 ± 0.2^{ab}
	S2	1.85 ± 0.2^{b}	1.67 ± 0.03^{b}	2.79 ± 0.09^{a}	1.55 ± 0.05^{b}
p Values					
Exper. site		<0.001	0.002	<0.001	0.005
Season		0.13	0.82	0.6	0.09
Site × season		0.37	0.018	0.4	0.9

Table 2. Land equivalent ratio (LER) for grain yield, total biomass (TDW), nitrogen (N) and phosphorus (P) uptake under S1 (P deficient) and S2 (P sufficient) conditions.

The complementarily of N and P use between cereals and N_2-fixing legumes, where the two species compete for the same soil's pool of N and P, the legume, through symbiotic N_2 fixation, can essentially access to the additional pool of atmospheric N_2 [10].

Facilitation occurs some species increases either growth or N-P nutrition of another species [25]. Recently, some research studies reported that advantage of both intercropping maize-common bean and maize-cowpea was confirmed for N and P acquisition by either chickpea or durum wheat [10].

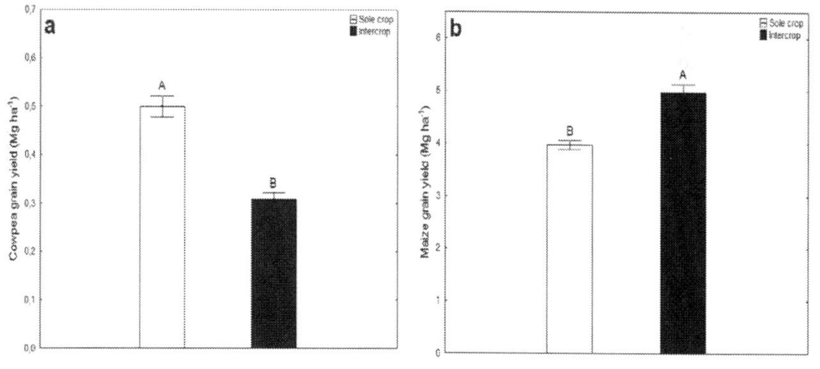

Figure 7. Grain yields (Mg ha^{-1}) of cowpea (a) and maize (b) in different cropping systems.

This intercropping advantage recorded more than 24% N uptake compared to sole crop. Similarly, under Mediterranean conditions with P-deficient soils, it was confirmed that maize grain yield in a maize-cowpea (**Figure 7**) [7] and common bean-cowpea intercropping under

P-low soil substantially increased (25%) compared to sole-cropped maize [7, 10]. Likewise, grain yield of either maize [26] or durum wheat [14] was increased when intercropped with cowpea and faba bean; respectively. In another study, above- and belowground interactions in a wheat-soybean intercropping differentially contributed (30% and 23%, respectively) in yield increase [27].

7. Conclusion

The main aim of this revision of the literature was to explain the effect of intercropping chickpea and durum wheat on N and P acquisition, especially under Mediterranean conditions.

Intercropping of cereal and legume can improve P and N use efficiency for crop growth and grain yield compared to sole crops. Enhanced soil P and N acquisition by root activity of either intercropped legume or cereal has been proposed as a mechanism of facilitation. It has also been reported that facilitation was more pounced by interspecific competition when P and N are more limiting for crop growth. Biomass, grain yield and consequently the taken up amount of N and P of intercropped cereals were significantly increased compared to those observed as sole crop. Presumably, pH change, increase in EURS and root respiration in legumes rhizosphere were the root-induced processes implied in the enhanced N and P availability for intercropped cereals. Indeed, in low P calcareous soils, the increased P availability can significantly improved aboveground of biomass in intercropping, though it mainly enhanced grain yield for intercropped cereals.

As conclusion, research findings reported in this present revision suggest that intercrops promote an advantage in grain yield and N-P nutrition for both cereal and legume. This legumes facilitation would have been related to root-induced changes modifying N and P bioavailability in the rhizosphere, as a result of enhancing in EURS in low P soils conditions.

Author details

M. Latati[1*], S. Benlahrech[1], M. Lazali[2], Tellah Sihem[1], G. Kaci[1], R. Takouachet[1], N. Alkama[3], F.Z. Hamdani[1], E.A. Hafnaoui[1], B. Belarbi[1], G. Ounane[1] and S.M. Ounane[1]

*Address all correspondence to: m.latati@yahoo.com

1 National High School of Agronomy (ENSA), Department of Plant Production, Laboratory of Integrative Improving of Plan Production, Avenue Hassane Badi, El Harrach, Algiers, Algeria

2 University of Djilali Bounaama Khemis Miliana, Route Theniet El Had, Soufay, Ain Defla, Algeria

3 University of Mouloud Mammeri, Agronomy Department, Tizi Ouzou, Algeria

References

[1] Dawson CJ, Hilton J (2011) Fertiliser availability in a resource-limited world: production and recycling of nitrogen and phosphorus. Food Policy 36: 14–22.

[2] Latati M, Pansu M, Drevon JJ, Ounane SM (2013) Advantage of intercropping maize (*Zea mays* L.) and common bean (*Phaseolus vulgaris* L.) on yield and nitrogen uptake in Northeast Algeria. IJRAS 01: 1–7.

[3] Li L, Tilman D, Lambers H, Zhang FS (2014) Plant diversity and overyielding: insights from belowground facilitation of intercropping in agriculture. New Phytol 203: 63–69.

[4] Tang X, Bernard L, Brauman A, Daufrense T, Deleporte P, Desclaux D, Souche G, Placella S, Hinsinger P (2014) Increase in microbial biomass and phosphorus availability in the rhizosphere of intercropped cereal and legumes under field conditions. Soil Biol Biochem 75: 86–93.

[5] Dyson T (1999) World food trends and prospects to 2025. Proc Natl Acad Sci 96: 5929–5936.

[6] Betencourt E, Duputel M, Colomb B, Desclaux D, Hinsinger P (2012) Intercropping promotes the ability of durum wheat and chickpea to increase rhizosphere phosphorus availability in a low P soil. Soil Biol Biochem 46: 21–33.

[7] Latati M, Blavet D, Alkama N, Laoufi H, Drevon JJ, Gérard F, Pansu M, Ounane SM (2014) The intercropping cowpea-maize improves soil phosphorus availability and maize yields in an alkaline soil. Plant Soil 85: 181–191.

[8] Li YY, Yu C, Cheng X, Li CJ, Sun JH, Zhang FS, Lambers H, Li L (2009) Intercropping alleviates the inhibitory effect of N fertilization on nodulation and symbiotic N_2 fixation of faba bean. Plant Soil 323: 295–308.

[9] Cong WF, Hoffland E, Li L, Six J, Sun JH, Bao XG, Zhang FS, Van Der Wep W (2014) Intercropping enhances soil carbon and nitrogen. Global Change Biology. 21: 1715–1726.

[10] Latati M, Bargaz A, Belarbi B, Lazali M, Benlahrech S, Tellaha S, Kaci G, Drevon JJ, Ounane SM (2016) The intercropping common bean with maize improves the rhizobial efficiency, resource use and grain yield under low phosphorus availability. Eur J Agron 72: 80–90.

[11] Hinsinger P, Betencourt E, Bernard L, Brauman A, Plassard C, Shen J, Tang X, Zhang F, (2011). P for two sharing a scarce resource e soil phosphorus acquisition in the rhizosphere of intercropped species. Plant Physiol 156: 1078–1086.

[12] Wang Z, Bao X, Li X, Jin X, Zhao J, Sun J, Christie P (2015) Intercropping maintains soil fertility in terms of chemical properties and enzyme activities on a timescale of one decade. Plant Soil 391: 265–282

[13] Alkama N, Bolou Bi Bolou E, Vailhe H, Roger L, Ounane SM, Drevon JJ (2009) Genotypic variability in P use efficiency for symbiotic nitrogen fixation is associated with variation of proton efflux in cowpea rhizosphere. Soil Biol Biochem 41: 1814–1823.

[14] Li H, Shen J, Zhang F, Clairotte M, Drevon JJ, Le Cadre E, Hinsinger P (2008). Dynamics of phosphorus fractions in the rhizosphere of common bean (*Phaseolus vulgaris* L.) and durum wheat (*Triticum turgidum durum* L.) grown in monocropping and intercropping systems. Plant Soil 312: 139–150.

[15] Devau N, Le Cadre E, Hinsinger P, Gérard F (2011a) Effects of inorganic fertilization and pH on processes and mechanisms controlling dissolved inorganic phosphorus in soils. Geochim Cosmochim Acta 75: 2980–2996.

[16] Alkama N, Ounane G, Drevon, JJ (2012) Is genotypic variation of H^+ efflux under P deficiency linked with nodulated-root respiration of N_2-fixing common-bean (*Phaseolus vulgaris* L.)? J Plant Physiol 169: 1084–1089.

[17] Zhang F, Li L (2003) Using competitive and facilitative interactions in intercropping systems enhances crop productivity and nutrient-use efficiency. Plant Soil 248: 305–312.

[18] Maingi MJ, Shisanya AC, Gitonga MN, Hornetz B (2001) Nitrogen fixation by common bean (*Phaseolus vulgaris* L.) in pure and mixed stands in semi arid South east Kenya. Eur J Agron 14: 1–12.

[19] Depret G, Laguerre G (2008) Plant phenology and genetic variability in root and nodule development strongly influence genetic structuring of *Rhizobium leguminosarum* biovar viciae populations noduling pea. New Phytol 179: 224–235.

[20] Bargaz A, Ghoulam C, Amenc L, Lazali M, Faghire M, Abadie J, Drevon JJ (2012) A phosphoenol pyruvate phosphatase transcript is induced in the rootnodule cortex of *Phaseolus vulgaris* under conditions of phosphorus deficiency. J Exp Bot 63: 4723–4730.

[21] Pan XW, Li WB, Zhang QY, Li YH, Liu MH (2008) Assessment on phosphorus efficiency characteristics of soybean genotypes in phosphorus-deficient soils. Agric Sci China 7: 958–969.

[22] Devau N, Hinsinger P, Le Cadre E, Gérard F (2011b) Root-induced processes controlling phosphate availability in soils with contrasted P-fertilized treatments. Plant Soil 348: 203–218.

[23] Li L, Tang C, Rengel Z, Zhang F (2003) Chickpea facilitates phosphorus uptake by intercropped wheat from an organic phosphorus source. Plant Soil 248: 297–303.

[24] Callaway RM (1995) Positive interactions among plants. Bot Rev 61: 306–349.

[25] Song YN, Zhang FS, Marschner P, Fan FL, Gao HM, Bao XG, Sun JH, Li L (2007) Effect of intercropping on crop yield and chemical and microbiological properties in rhizosphere of wheat (*Triticum aestivum* L.), maize (*Zea mays* L.), and faba bean (*Vicia faba* L.). Biol Fert Soil 43: 565–574.

[26] Dahmardeh M, Ghanbari A, Syahsar BA, Ramrodi M (2010) The role of intercropping maize (*Zea mays* L.) and Cowpea (*Vigna unguiculata* L.) on yield and soil chemical properties. African J Agric Res 5 (8): 631–636.

[27] Zhang F, Zhang S, Zhang J, Zhang R, Li F (2004) Nitrogen fertilization on uptake of soil inorganic phosphorus fractions in the wheat root zone. Soil Sci Soc America J 68: 1890–1895.

Chapter 7

Grain Legume Consumption Inhibits Colorectal Tumorigenesis: A Meta-Analysis of Human and Animal Studies

Thushanthi Perera , Yumie Takata and Gerd Bobe

Additional information is available at the end of the chapter

http://dx.doi.org/10.5772/63099

Abstract

Grain legume consumption has been linked in meta-analysis studies to decreased risk of metabolic syndrome, obesity, and cardiovascular diseases; however, the evidence for a chemo-protective effect of grain legume consumption against colorectal tumorigenesis has been considered inconclusive. We conducted a meta-analysis of human and animal studies to evaluate the effect of grain legume consumption on colorectal cancer (CRC) and its precursors. Twelve case-control studies (42,473 controls and 12,408 cases) and 11 prospective cohorts (1,533,527 participants including 12,274 cases) were included in the meta-analysis; the pooled risk ratio (95% confidence interval) for the highest versus the lowest legume intake group based on a random effects model was 0.72 (0.60–0.89) for incident adenoma, 0.91 (0.84–0.99) for prevalent adenoma, and 0.82 (0.74–0.91) for CRC. Fourteen animal studies (355 animals on grain legume diets and 253 animals on control diets) were included in the meta-analysis and showed in all but one study a chemo-preventive effect against colorectal tumorigenesis. Grain legumes contain various compounds (e.g., resistant starch, soluble fiber, insoluble fiber, phytosterols, saponins, phytates, flavonoids, proanthocyanidins, and phenolic acids) that have been shown to inhibit colorectal tumorigenesis in animal studies at concentrations that are relevant for human diets. Grain legume consumption alters several molecular pathways (e.g., p53, mTOR, NF-kB, Akt, and AMPK) that are critical for tumor induction, promotion, and progression. Based on our meta-analysis, daily grain legume consumption confers chemo-preventive effects against CRC.

Keywords: grain legumes, colorectal cancer, meta-analyses, bioactive compounds, molecular mechanisms

1. Introduction

Grain legumes (i.e., pulses) are defined as plants belonging botanically to the family *Leguminosae,* which are harvested as dry seeds for food consumption [1–3]. Grain legumes are behind cereal grains the most common food crop worldwide; the primarily grown grain legumes are in the order as follows: dry beans (e.g., pinto, navy, red kidney, lima, butter, white, and black beans; *Phaseolus and Vigna* ssp.), chick peas (i.e., garbanzo beans; *Cicer arietinum*), dry peas (e.g., garden peas; *Pisum sativum*), dry cow peas (*Vigna unguiculata*), lentils (*Lens culinaris*), and dry broad beans (e.g., horse beans; *Vicia faba*) [3–5]. Beans are oval or kidney shaped, peas are round, and lentils are flat. Grain legumes have served as staple foods in many cultures around the globe, as they can be grown relatively inexpensively in various climate zones and have a health-promoting nutrient profile, that is, they are a good dietary source of protein, rich in fiber and folate, and very low in saturated fatty acids, cholesterol, and sodium [6–8].

Grain legume consumption dramatically decreased in westernizing countries [9] and is in the U.S., similar to other Western countries [10, 11], on average low (12.9 g/d) and infrequent (only 8 and 14% consumed grain legumes daily or every other day) [6, 12]. Given the health-promoting properties and nutrient profile of grain legumes and the growing interest in ethnic, gluten-free, and vegetarian cuisine in Western countries, increasing grain legume consumption represents an important public health opportunity for chronic disease prevention.

A research focus is the use of legumes for cancer prevention, specifically colorectal cancer (CRC) [4]. Globally, CRC is the third most common cancer in men and the second most common in women [13]. Two recent A meta-analysis study reported a protective effect of legume consumption for colorectal adenomas (CRAs) in case-control and cohort studies (combined odds ratio (OR) = 0.83; 95% confidence interval (CI): 0.75–0.93) and CRC in cohort studies (OR = 0.91; 95% CI: 0.84–0.98) [14, 15]. Both meta-analysis studies, however, included studies in which participants consumed legumes primarily as soy products (i.e., studies conducted in China, Japan, Malaysia, and South Korea), as opposed to grain legumes (i.e., studies conducted in Africa, North and South America, and Europe). Moreover, the meta-analysis of CRC showed a protective effect for soybeans (OR = 0.85; 95% CI: 0.73–0.99) but not for other beans (OR = 1.00; 95% CI: 0.89–1.13) [15]. A third meta-analysis study published in 2010 reported no statistically significant association between legume fiber consumption and CRC in four prospective U.S. and European studies combined (OR = 0.89; 95% CI: 0.78–1.02) [16].

The objective of this chapter is to evaluate the evidence of a chemo-preventive role of grain legume consumption in colorectal tumorigenesis in human (ecological, case-control, and cohort studies) and animal studies by conducting a literature review and meta-analyses. The goal is to suggest areas of future research and provide up-to-date scientific evidence for dietary recommendation of legume consumption.

2. Colorectal cancer: incidence, mortality, and risk factors

Worldwide, annually 1.361 million new CRC cases and 0.694 million deaths due to CRC accrue, according to GLOBOCAN in 2012 [13, 17]. In the U.S., the lifetime risk of being diagnosed with CRC is 5% and the treatment costs were estimated to be over $14 billion [18, 19], highlighting CRC prevention as a public health priority. CRC development is a multistep process over many years, often decades, involving usually random genetic mutations in colorectal epithelial cells causing the activation of tumor-promoting genes and the loss of tumor suppressor gene function [20, 21]. Starting often as aberrant crypt foci (ACF), most CRC arise from benign, adenomatous polyps (i.e., adenomas) that grow from glandular cells of the colorectal epithelial lining into advanced adenomas and then adenocarcinomas [22–24]. Over 50% of the Western population will develop colorectal adenomas (CRAs) by the age of 70 [23]. Less than 10% of adenomas, however, progress to become invasive and spread to adjacent blood or lymph vessels [25]. Success of CRC treatment depends on early detection. If CRC has not spread beyond the colorectal wall (i.e., localized stage), 5-year survival rates are 90.3%; however, survival rates decline when CRC has spread to lymph nodes and/or nearby tissue (i.e., regional disease; a 5-year survival of 70.4%) and are low when CRC has spread to other organs (i.e., distant disease; a 5-year survival of 12.5%) [26]. Currently, only 40% of CRC patients are diagnosed with localized stage, highlighting that importance of early detection and treatment of CRC and its precursors [27].

Genetics is an important CRC risk factor. About 20% of CRC patients have a family history of CRC (10–15% lifetime risk for patients with one first-degree relative; 20% lifetime risk for patients with at least two first-degree relatives or one first-degree relative diagnosed with CRC before age 45) and 2–4% have a well-defined genetic syndrome (i.e., Lynch syndrome and familial polyposis; 80–90% lifetime risk) [19]. Chronic inflammation, specifically inflammatory bowel disease (IBD), is another important CRC risk factor with a 10–20% lifetime risk, which is increased among patients with a longer IBD history [19, 28]. Other important medical CRC risk factors are obesity, insulin resistance, and diabetes mellitus; CRC risk increases linearly with duration and severity of those morbidities [19, 29–33]. Modifiable CRC risk factors include smoking, heavy alcohol consumption, and sedentary behavior, each with a 6% lifetime risk [19], whereas medications such as aspirin and nonsteroidal anti-inflammatory drugs (NSAIDs) and hormone-replacement therapy in postmenopausal women can decrease CRC risk [19].

Food and nutrition play an important part in the etiology and prevention of CRC and may account for 70–90% of all cases [34–36]. A panel of experts, primarily epidemiologists organized by the World Cancer Research Fund (WCRF) and the American Institute for Cancer Research (AICR), evaluated the scientific evidence on food, nutrition, and physical activity on cancer risk [34]. Human studies were ordered according to the quality of the study design as follows: (1) ecological studies (lowest quality; most susceptible to confounders; i.e., factors that are associated with both disease status and the evaluated food); (2) case-control studies (very susceptible to recall bias; i.e., selective reporting of the diet after disease diagnosis); (3) Prospective cohort studies; and (4) clinical trials (gold standard and least susceptible to bias). In the case of substantial amount of evidence available, the panel focused on studies using high-quality designs. Evidence from animal and cell culture studies was taken into account to demonstrate plausible mechanism for diet and cancer association. Based on the evidence, an individual food, food group, or individual nutrient was classified for each cancer site as "convincing", "probable", "limited-suggestive", or "limited-no conclusion" decreases the risk or increases the risk [34].

In 2007, the panel classified red and processed meat consumption as convincingly increases CRC risk, whereas calcium and foods containing fiber were classified as probably decreases CRC risk, and selenium and foods containing folate were classified as limited-suggestive evidence for decreasing CRC risk [34]. No conclusion was made for legumes and CRC risk because of the limited data available in 2007 [34]. As in the last 8 years more data have been collected, we reevaluate in this chapter the evidence on the relation between grain legume consumption and CRC risk. We hypothesized a protective effect of grain legume consumption on CRC risk because grain legumes are an excellent dietary source of fiber (5.7–9.0 g/100 g of cooked legumes) and folate (83–174 μg/100 g of cooked legumes) [7], both of which were classified as decreasing CRC risk in 2007 [34].

3. Grain legumes and colorectal neoplasia in human studies

Ecological studies examine the association between diet and disease on the population level; five studies evaluated the relation between legume intake and risk of CRC incidence or mortality on the population level and observed inconsistent relations [9, 37–40]. Correa reported that countries with higher bean consumption in 1964–1966 had lower colon cancer mortality rates 7–9 years later ($r = -0.68$) [40]. Similarly, Bejar *et al.* stated that the decrease in legume consumption between 1960 and 1990 coincided with an increase in CRC incidence and mortality rates 10 years later in Spain [37, 39]. In follow-up studies, Bejar *et al.* extended the analysis to 15 European and 13 non-European countries [9, 38]. Whereas the strong inverse relation between legume intake and CRC incidence rates held true for some countries (i.e., Norway, Spain, Germany, and France), other countries (i.e., Australia, Italy, and Colombia) had positive relations, as a result of a slight increase in legume consumption between 1965 and 2005. Thus, changes in legume consumption alone cannot explain the temporal changes in CRC incidence rates; rather, changes toward a Western diet were associated with an increased CRC risk (depending on country of origin, adoption of a westernized diet either increased or decreased grain legume consumption). In support, Monroe *et al.* reported in a migrant study that an increase of CRC incidence rates (men: 85%; women: 95%) coincided with a 46% decrease in dry bean or pea consumption (57.0–26.6 g/d) from first- to second-generation Mexico-born U.S. Americans in the Multiethnic Cohort Study [41], and Haentzel *et al.* showed a detrimental effect of grain legume consumption on CRC incidence in Japan-born Hawaiian [42].

In case-control studies, participants with (cases) or without (controls) a disease recall their diet. Besides recalling a diet from past years, participants try to make sense of their disease outcome based on their lifestyle choices. Thus, foods and food groups that have been known to be associated with disease outcomes by the public are often erroneously associated with the disease outcome (i.e., selective reporting bias). Nineteen peer-reviewed publications (46,769 controls and 14,567 cases; two studies had each two publications [43, 44] and [45, 46]) evaluated in 17 case-control studies the relation between legume consumption and colorectal neoplasia; six studies reported prevalent adenomas as endpoint [47–52] and 11 studies reported carcinomas as endpoint [42–46, 53–60] (**Table 1**). Most case-control studies were from the U.S. ($n = 8$), five were from Europe, two were from South America, and one each from Australia and Jordan. Risk estimates specific to the intake of legumes (including soybeans and their products), grain legumes, and grain legume fiber were reported in six, 11, and two case-control studies, respectively. Gender-specific risk estimates were reported in five case-control studies, and cancer-site–specific risk estimates were reported for colon and rectum in seven and four case-control studies, respectively. Half of the studies showed a protective legume effect on CRA (**Table 1**), one of which was statistically significant [50]. A distinct clustering was observed for CRC. Seven of 11 case-control studies had significant risk estimates of 0.5 or lower [45, 46, 50, 53, 55, 56, 59, 60]; three of the six low-risk estimates were from women and, for the remaining three, no gender-specific risk estimates were reported. In contrast, the risk estimates of the other studies were around 0.9 (**Table 1**).

Reference, region (country)	Study period	Study design, no. controls/cases	Sex, age	Diet assessment	Grain legume, quantity for comparison, risk estimates (95% CI)	Matching/adjusting for confounders
Prevalent colorectal adenoma						
Sandler et al., 1993 [47] North Carolina (U.S.)	1988–1990	Colonoscopy Cases: 236 Controls: 409	Both, ≥30 years, no CRC, IBD history	Phone interview: FFQ with >100 food items for previous yr	Grain legume fiber Men: ≥3.14 vs. <0.97 g/d OR = 0.99 (0.43–2.29) Women: ≥2.17 vs. <0.61 g/d OR = 1.26 (0.63–2.51)	No matching specified Adjusted for age, alcohol intake, BMI, and total energy intake
Witte et al., 1996 [48] California (U.S.)	1991–1993	Sigmoidoscopy Cases: 488 Controls: 488	Both 50–74 years; no CRA, IBD history	Personal interview: FFQ with 126 food items for previous yr	Legumes (beans, lentils, peas, lima beans, tofu, soybeans, peanut butter) Mean 8.5 vs. 0.5 servings/wk OR: 0.85 (0.56–1.28)	Matched by age, sex, day of sigmoidoscopy, Kaiser center Adjusted by race, BMI, physical activity, smoking, and intake of total energy and saturated fat
Smith-Warner et al., 2002 [49] Minnesota Cancer Prevention Research Unit Study (U.S.)	1991–1994	Colonoscopy and population Cases: 564 Controls: 682 colonoscopy, 535 population	Both, 30–74 years, no CRA, IBD history	Self-administered FFQ precolonoscopy with >153 food items for previous yr	Legumes (alfalfa sprouts, beans, peas) Men: Mean 5.0 vs. 1.0 servings/wk Colonoscopy: OR = 0.96 (0.62–1.49) Population: OR = 1.15 (0.77–1.72) Women: Mean 5.5 vs. 1.1 servings/wk Colonoscopy: OR = 1.08 (0.68–1.74) Population: OR = 0.96 (0.58–1.59)	Matched by age, sex, and residence Adjusted for age, total energy and fat intake, BMI, smoking, alcohol, NSAID use, multivitamin use, and hormone replacement therapy
Agurs-Collins et al., 2006 [50] African-American (U.S.)	2001–2003	Colonoscopy Cases: 53 Controls: 133	Both, 29–81 years	FFQ with 39 food items (Rate Your Diet Quiz)	Grain legumes (dry beans, split peas, lentils) ≥3× vs. ≤1×/wk OR = 0.19 (0.04–0.91)	No matching specified Adjusted for age, smoking, alcohol, sex, weight, aspirin use, alcohol, family CRC history, and exercise
Millen et al., 2007 [51] Prostate, Lung, Colorectal, and Ovarian Cancer Screening Trial (PLCO)	1993–2001	Sigmoidoscopy Cases: 3057 Controls: 29,413	Both, 55–74 years; no CRA, IBD history	Self-administered FFQ pre-, on, or post-sigmoidoscopy with 137 food items for previous yr	Legumes (beans, peas, tofu, and soybeans) Median 0.4 vs. 0.05 energy-adjusted servings/wk OR = 0.92 (0.81–1.03) Sex and age adjusted: OR = 0.85 (0.75–0.96)	Matching not specified Adjusted for age, sex, race, education, family CRC history, smoking, alcohol use, aspirin use, replacement hormone use, physical activity, BMI
Wu et al., 2009 [52] Tennessee Colorectal Polyp Study (U.S.)	2003–2005	Colonoscopy Cases: 764 Controls: 1517	Both, 40–75 years, no CRA, IBD history	FFQ with >108 food items for previous yr	Grain legumes (green beans and peas, dry and canned beans) Tertile T3 vs. T2 Quantity N/A OR = 0.95 (0.74–1.24)	No matching specified Adjusted for age, sex, race, study location, BMI, smoking, alcohol consumption, NSAID use, physical activity, education level, family income, family CRC history, and intake or total energy and red meat

Reference, region (country)	Study period	Study design, no. controls/cases	Sex, age	Diet assessment	Grain legume, quantity for comparison, risk estimates (95% CI)	Matching/adjusting for confounders
Colon Cancer						
Iscovich et al., 1992 [56] La Plata (Argentina)	1985–1987	Population Cases: 110 Controls: 220	Both, 35–80 years	Personal interview FFQ with 140 food items for previous 5 yrs	Grain legumes (beans, lentils, peas, and chick peas) Quartile 4 vs. 1 OR = 0.52 (0.24–1.12) Quartile 3 vs. 1 OR = 0.32 (0.14–0.73) Quantity N/A	Matched by age and gender Adjustment not specified
Steinmetz et al., 1993 [60] Adelaide (Australia)	1979–1980	Population Cases: 220 Controls: 438	Both, 30–74 years	Self-administered FFQ with 141 food items a yr ago	Legumes (green, dry and broad beans, lentils, dry and chick peas, and soybeans) Men: >1 vs. 0 servings/wk OR = 0.74 (0.38–1.45); Women: >0.6 vs. 0 servings/wk OR = 0.43 (0.20–0.93)	Matched by age and gender Adjusted for protein intake, occupation, Quetelet's index, alcohol consumption, and age at first live birth (only women)
Kampman et al., 1995 [57] (Netherlands)	1989–1993	Population Cases: 232 Controls: 259	Both, ≤75 years, no history of CR tumors	Personal interview: FFQ with 289 food items for previous yr	Legumes Quartile 4 vs. 1 (infrequent legume consumption) OR = 1.08 (0.67–1.76)	Matched by age, gender, and degree of urbanization Adjustment not specified
Colorectal Cancer						
Haenszel et al., 1973 [42] Hawaiian born in Japan (U.S.)	1966–1970	Hospital Cases: 179 Controls: 357	Both Age N/A	Personal interview: four legumes, soybeans excluded	Grain legumes (green and red beans, peas, and Chinese peas) >21× vs. <8×/mo legumes RR = 3.5 95% CI N/A	Matched by sex and birth place Adjustment not specified
La Vecchia et al., 1988 [58] Milan (Northern Italy)	1985–1987	Hospital Cases: 339 colon, 236 rectal Controls: 778	Both, <75 years	Personal interview: 29 food items prior to disease diagnosis	Grain legumes Tertile 3 vs. 1 Quantity N/A Colon: RR = 1.04; Rectum: RR = 0.94 95% CI N/A	Matching not specified Adjusted for social class, age, sex, and area of residence
Benito et al., 1991 [53] Majorca (Spain)	1984–1988	Population and Hospital Cases: 286 Controls: 203 hospital 286 population	Both, <80 years	Personal interview: FFQ with 99 food items for previous yr	Grain legume fiber Quartile 4 vs. 1 Quantity N/A RR = 0.40 95% CI N/A	Matched by age and gender Adjusted for age, sex, body weight, and total energy intake
Bidoli et al., 1992 [54] Pordenone (North Eastern Italy)	1986–1990	Hospital Cases: 123 colon, 125 rectal Controls: 699	Both Age not specified	Personal interview: FFQ (number of food items N/A before disease) onset	Grain legumes Tertile 3 vs. 1 Quantity N/A Colon: RR = 1.2 Rectum: RR = 0.8 95% CI N/A	Matched by hospital Adjusted for age, gender, and social status
Le Marchand et al., 1997 [59] Hawaii Multiethnic (U.S.)	1987–1991	Population Cases: 1192 Controls: 1192	Both <85 years, no history of colorectal tumors	Personal interview: FFQ with 282 food items 3 yrs before disease onset	Legumes (including soy products) Men: >46 vs. <11 g/d OR = 0.8 (0.5–1.2) Women: >44 vs. <9 g/d OR = 0.5 (0.3–0.9)	Matched by age, sex, and race Adjusted for age, family CRC history, alcohol consumption, smoking, physical activity, Quetelet index, and intake of total calories, eggs, and calcium
Franceschi et al., 1998 [55] (Italy)	1991–1996	Hospital Cases: 1225 colon 728 rectal Controls: 5155	Both Age not specified	Personal interview: FFQ with 98 food items 2 yrs before disease diagnosis	Grain legumes (beans and peas) >3 vs. <0.5 servings/wk Colon: OR = 0.5 (0.4–0.7) Rectum: OR = 0.7 (0.5–0.9)	Matching not specified Adjusted for age, sex, center, year of interview, education, physical activity, alcohol consumption, and total energy intake
Deneo-Pellegrini et al., 2002 [46] Montevideo (Uruguay)	1996–2002	Hospital Cases: 484 Controls: 1452	Both 30–89 years	Personal interview: FFQ with 64 food items a yr before disease diagnosis	Grain legumes (kidney beans and lentils) Quartile 4 vs. 1 Quantity N/A Overall: OR = 0.7 (0.5–0.9) Men: OR = 0.8 (0.5–1.2) Women: OR = 0.5 (0.3–0.9) Colon: OR = 0.9 (0.9–1.1) Rectum: OR = 0.8 (0.7–0.9)	Matched on age, sex, residence, and urban/rural status Adjusted for age, sex, rural/urban status, education, first-degree family CRC history, BMI, and intake of total energy and red meat
Aune et al., 2009 [45] Montevideo (Uruguay)	1996–2004	Hospital Cases: 3539 Controls: 2032	Both 26–89 years	Personal interview: FFQ with 64 food items a yr before disease diagnosis	Grain legumes (kidney beans and lentils) Legume: Median 14.38 vs. 1.35 g/d OR = 0.43 (0.32–0.59) Beans: Median 9.44 vs. 0 g/d OR = 0.44 (0.31–0.61) Lentils: Median 11.68 vs. 0 g/d OR = 0.53 (0.38–0.75)	Matching not specified Adjusted for age, sex, residence, BMI, education, income, interviewer, smoking status and history, alcohol consumption, mate drinking status, and intake of total energy, dairy products, fatty foods (eggs, cake,

http://dx.doi.org/10.5772/63099

Reference, region (country)	Study period	Study design, no. controls/cases	Sex, age	Diet assessment	Grain legume, quantity for comparison, risk estimates (95% CI)	Matching/adjusting for confounders
						custard, butter), fruits and vegetables, and total meat
Abu Mweis et al., 2015 [43] (Jordan)	2010–2012	Hospital Cases: 167 Controls: 240	Both >18 years	Self-administered FFQ with 109 food and beverage items (DHQ 1) a yr before disease diagnosis	Lentils >1× vs. <1×/wk OR = 1.49 (0.80–2.79)	Matched by age, sex, occupation, and marital status Adjusted for age, sex, family CRC history, physical activity, smoking, education level, marital status, work, income, and total energy intake
Tayyem et al., 2015 [44] (Jordan)	2010–2012	Hospital Cases: 220 Controls: 281	Both >18 years	Self-administered FFQ with 109 food and beverage items (DHQ 1) a yr before disease diagnosis	Lentils 1×/wk vs. <1×/mo OR = 1.3 (0.72–2.4) White beans 1×/wk vs. <1×/mo OR = 0.86 (0.37–2.1) Green beans 1×/wk vs. <1×/mo OR = 1.0 (0.57–2.2) Peas 1×/wk vs. <1×/mo OR = 1.0 (0.44–2.0)	Matched by age, sex, occupation, and marital status Adjusted for age, sex, family CRC history, physical activity, smoking, education level, marital status, work, income, and total energy intake

*Statistically significant association of legume consumption and colorectal neoplasia.
CRA: colorectal adenoma; CRC: colorectal cancer; FFQ: food frequency questionnaire; IBD: inflammatory bowel disease; mo: month; N/A: not available; OR: odds ratio; RR: relative risk; wk: week; 95% CI: 95% confidence interval.
1 serving of legume equals 0.5 cup of cooked legumes (~90 g) [7].

Table 1. Description of retrospective case-control studies of grain legume consumption and colorectal neoplasia.

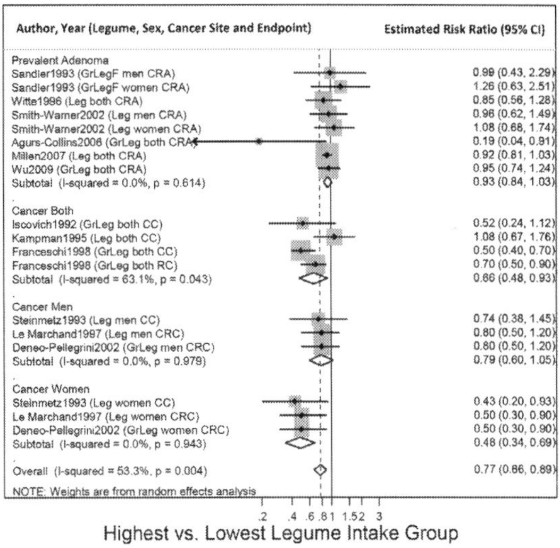

Figure 1. Forest plot of legume consumption (highest vs. lowest category) and colorectal neoplasia risk in retrospective studies stratified by type of neoplastic lesion and gender (only for cancer studies). The dot in each study indicates the estimated risk ratio, vertical bars represent 95% CI, and the size of gray square box reflects the study's weight in the random effects meta-analysis. The straight line indicates no association and the dashed line indicates the summary risk estimate across all studies. The open diamond on the bottom indicates the pooled risk estimate and the right vertices of the diamond reflect the 95% CI. CC: colon cancer; RC: rectal cancer; CRC: colorectal cancer; CRA: colorectal adenoma; GrLeg: grain legume; GrLegF: grain legume fiber; Leg: legume; LegF: legume fiber.

Meta-analysis using a random effects model of natural log odds ratios (OR) in STATA was possible for 12 case-control studies [46–52, 55–57, 59, 60] that included 12,408 cases and 42,473 controls. We had to exclude the four oldest case-control studies [42, 53, 54, 58] because the 95% CIs were not reported and two case-control studies [43, 44] provided only estimates of individual grain legumes. We checked for heterogeneity of estimates, influential risk estimates, and publication bias using funnel plots and Egger's method. When comparing the highest versus the lowest legume intake group, we observed a protective effect of legume consumption on CRA (relative risk (RR) = 0.93; 95% CI: 0.84–1.03; P = 0.15) and CRC (RR = 0.65; 95% CI: 0.54–0.77; P <0.001). There was moderate heterogeneity (30.2%) among studies for CRC risk (P = 0.17), but <0.01% for CRA risk. The range of risk estimates was 0.56–0.65 for CRC after removing one study at a time. No significant publication bias was observed (P = 0.11). The heterogeneity among CRC risk estimates could be explained by gender-specific differential dietary recalls (**Figure 1**); the protective effect of legume consumption on CRA was in men, RR = 0.79 (95% CI: 0.84–1.03; P = 0.10; <0.01 heterogeneity), in women, RR = 0.49 (95% CI: 0.34–0.69; P <0.0001; <0.01% heterogeneity), and intermediate RR = 0.67 (95% CI: 0.48–0.93; P = 0.02; 63.1% heterogeneity) in studies that did not provide gender-specific estimates.

In prospective cohort studies, dietary information of cohorts or groups of healthy individuals at the time of study recruitment is linked to subsequent disease outcomes. We evaluated the relation between legume consumption and colorectal neoplasia in 15 peer-reviewed publications from 11 prospective cohorts (1,621,519 participants with 13,546 cases), 11 reported cancer as endpoint [61–71] and the remaining four studies reported incident and/or prevalent adenomas as endpoint [72–75] (**Table 2**). All, except for two European cohorts, were U.S. cohorts. Risk estimates were reported for men in six and for women in eight prospective cohorts. Risk estimates specific to colon and rectum were reported in two and one cohorts, respectively. Risk estimates specific to the intake of legumes, legume fiber, grain legumes, and grain legume fiber were reported in three, three, three, and two cohorts, respectively. Two cohorts (Adventist Health Study and Polyp Prevention Trial) showed significant protective effects of grain legume consumption [69, 72, 75]. Four cohorts (Breast Cancer Detection Demonstration Project, Women's Health Study, Multiethnic Cohort Study, and NIH-AARP Study) showed a protective effect on CRC risk, the effect was statistically significant in some statistical models in the latter three cohorts [63, 64, 66–68]. Two cohorts (Nurses' Health Study and Health Professionals' Follow-up Study) showed a protective effect of legume consumption for CRA only [65, 73, 74]. Only three of the 11 cohorts (Iowa Women's Health Study and two European cohorts) showed no effects of legume consumption on CRC risk [61, 62, 70, 71].

Reference, cohort, country	Follow-up period	Study size, case no.	Sex, age	Diet assessment	Grain legume, quantity for comparison, risk estimates (95% CI)	Adjustment for confounders
Incident colorectal adenoma						
Lanza et al., 2006 [72]	1991–1994; 4-yr trial;	1905, 629	Both, >35 years	Four annual self-	Grain legumes (dry beans and lentils)	Adjusted for age, NSAIDs, sex,

Reference, cohort, country	Follow-up period	Study size, case no.	Sex, age	Diet assessment	Grain legume, quantity for comparison, risk estimates (95% CI)	Adjustment for confounders
U.S., Polyp Prevention Trial (PPT)	incident CRA <3 yrs old	No CRC, IBD history		administered FFQ with 27 food items and one grain legume question for previous yr	Mean: 45.1 vs. 3.1 g/d Any: OR = 0.78 (0.58–1.04) Men: OR = 0.69 (0.48–0.99) Advanced: OR = 0.30 (0.15–0.60)	intervention group, and sex by intervention group
Michels et al., 2006 [73] U.S., Nurses' Health Study (NHS)	Diet: 1980–1994, incident CRA >2 yrs old	9735, 633 No CRA, IBD history	Women 30–55 years in 1976	Self-administered FFQ with 61 food items for previous yr	Legumes (beans, lentils, peas, lima beans, tofu, soybeans) ≥5 vs. ≤ 1 serving/wk New Incidence only: OR = 0.67 (0.51–0.90) Trend: OR = 0.92 (0.87–0.98)	Adjusted for age, family CRC history, height, BMI, regular vigorous exercise, regular aspirin use, pack-years of smoking, current multivitamin supplement use, alcohol consumption, menopausal status, postmenopausal hormone use, and intake of total energy, red meat, and calcium
Tantamango et al., 2011 [75] U.S., Adventist Health Study (AHS)	Diet: 1976–1977, Endpoint: 2002–2004 incident CRA <20 yrs old	2818, 441 No CRC, IBD history	Both, All underwent colonoscopy, no age exclusion	Self-administered FFQ with 55 food and beverage items	Grain legumes (beans, lentils, split peas) ≥3×/wk vs. <1×/mo OR = 0.67 (0.44–1.01) Trend: $P = 0.02$	Adjusted for age, sex, education, BMI, and red meat intake
Prevalent colorectal adenoma						
Platz et al., 1997 [74] U.S., Health and Professionals' Follow-up Study (HPFS)	1986–1994	16,448, 690 No CRA, IBD history	Men 40–75 years All underwent colonoscopy	Self-administered FFQ with 127 food items for previous yr	Legume fiber (beans, lentils, peas, lima beans, tofu, soybeans) Median 2.6 vs. 0.5 g/d	Adjusted for age, family CRC history, prior endoscopy, BMI, smoking, multivitamin use, physical activity,

Reference, cohort, country	Follow-up period	Study size, case no.	Sex, age	Diet assessment	Grain legume, quantity for comparison, risk estimates (95% CI)	Adjustment for confounders
					RR = 0.82 (0.60–1.11) Trend: P = 0.06	regular aspirin use, and intake of energy, alcohol, red meat, folate, and methionine
Michels et al., 2006 [73] U.S., Nurses' Health Study (NHS)	Diet: 1980–1994 Endpoint: 1980–1998	34,467, 1720 No CRC and IBD history	Women 30–55 years in 1976	Self-administered FFQ with 61 food items for previous yr	Legumes (beans, lentils, peas, lima beans, tofu, soybeans) ≥5 vs. ≤1 serving/wk OR = 0.89 (0.75–1.05) Trend: OR = 0.96 (0.93–1.00)	Adjusted for age, family CRC history, height, BMI, regular vigorous exercise, regular aspirin use, pack-years of smoking, current multivitamin supplement use, alcohol consumption, menopausal status, postmenopausal hormone use, and intake of total energy, red meat, and calcium
Colon cancer						
Steinmetz et al., 1994 [70] U.S., Iowa Women's Health Study (IWHS)	1986–1990	41,837, 212	Women, 55–69 years at baseline, no CRC history	Self-administered FFQ with 127 food items for previous yr	Legumes (beans, lentils, peas, lima beans, tofu, soybeans) ≥1.0 vs. 0 servings/wk RR = 0.95 (0.66–1.36)	Adjust for age and total energy intake
Singh and Fraser, 1998 [69] U.S., Adventist Health Study (AHS)	1976–1982	32,051 157 Non-hispanic white	Both >25 years	Self-administered FFQ with 55 food items	Grain legumes (beans, lentils, split peas) >2× vs. <1×/wk RR = 0.53 (0.33–0.86)	Adjusted for age, sex, BMI, physical activity, parental CRC history, smoking, alcohol consumption, and aspirin use

Reference, cohort, country	Follow-up period	Study size, case no.	Sex, age	Diet assessment	Grain legume, quantity for comparison, risk estimates (95% CI)	Adjustment for confounders
Colorectal cancer						
Michels et al., 2000 [65] U.S., Nurses' Health Study (NHS)	1980–1996	88,764, 724	Women 30–55 years	Self-administered FFQ with 127 food items for previous yr	Legumes (beans, lentils, peas, lima beans, tofu, soybeans) ≥4 vs. <1 serving/wk RR = 1.26 95% CI N/A RR = 1.49 (1.04–2.12) per additional serving/wk	Adjusted for age, family CRC history, sigmoidoscopy, height, BMI, pack-years of smoking, alcohol consumption, physical activity, intake of total energy and red meat, and use of menopausal hormones, aspirin, and vitamin supplements
Michels et al., 2000 [65] U.S., Health and Professionals' Follow-up Study (HPFS)	1986–1996	47,325, 457	Men 40–75 years	Self-administered FFQ with 127 food items for previous yr	Legumes (beans, lentils, peas, lima beans, tofu, soybeans) ≥4 vs. <1 serving/wk RR = 0.97 95% CI N/A RR = 0.90 (0.57–1.42) per additional serving/wk	Adjusted for age, family CRC history, sigmoidoscopy, height, BMI, pack-years of smoking, alcohol consumption, physical activity, intake of total energy and red meat, and use of menopausal hormones, aspirin, and vitamin supplements
Voorrips et al., 2000 [71] Netherlands Cohort Study on Diet and Cancer (NCSDC)	1986–1992	Male: 58,279, 514 Women: 62,573, 396	Both, 55–69 years	Self-administered FFQ with 155 food items for previous yr	Grain legumes (green and lima beans) Male: Median 62 vs. 11 g/d Colon RR = 1.13 (0.77, 1.64) Rectum: RR = 0.92 (0.58–1.47) Female: Median 58 vs. 11 g/d	Adjusted for age, alcohol consumption, and family CRC history

Reference, cohort, country	Follow-up period	Study size, case no.	Sex, age	Diet assessment	Grain legume, quantity for comparison, risk estimates (95% CI)	Adjustment for confounders
					Colon RR = 0.79 (0.52, 1.20) Rectum: RR = 1.01 (0.53–1.94)	
Mai et al., 2003 [64] U.S., Breast Cancer Detection Demonstration Project (BCDDP)	1987–1998	45,491, 487	Women Age range N/A	Self-administered FFQ with 62 food items for previous yr	Grain legume fiber >1.38 vs. <0.20 g/1000 kcal/d RR = 0.84 (0.63–1.10)	Unadjusted
Bingham et al., 2003 [61] 10 EU countries, EPIC	1992–2002	519,978, 1065	Both, 35–70 years	Country-specific FFQ with 300–350 food items	Legume fiber Mean: 1.73 vs. 0.45 g/d HR = 1.04 (0.84–1.30)	Adjusted for age, weight, height, sex, intake of nonfat and fat energy, and stratified by center
Bingham et al., 2005 [62] 10 EU countries, EPIC	1992–2004	519,978, 1721	Both, 35–70 years	Country-specific FFQ with 300–350 food items	Legume fiber Mean: 1.9/1.0 vs. 0 g/d HR = 0.94 (0.79–1.14)	Adjusted for age, weight, height, sex, intake of nonfat and fat energy, and stratified by center
Lin et al., 2005 [63] U.S., Women's Health Study (WHS)	1993–2003	39,876, 223	Women ≥45 years	Self-administered FFQ with 131 food items for previous yr	Legumes (dry beans, lentils, peas, lima and green beans, tofu, soybeans) Median 0.9 vs. 0.1 serving/d RR = 0.83 (0.54–1.28) Legume fiber Median 1.8 vs. 0.4 g/d RR = 0.60 (0.40–0.91)	Adjusted for age, randomized treatment assignment, BMI, first-degree CRC family history, colon polyp history, physical activity, smoking status, baseline use of aspirin, hormone replacements, menopausal status, alcohol consumption, and intake of total energy and red meat
Nomura et al., 2007 [66] U.S., Multiethnic	1993–2001	191,011, 2110	Both, 45–75 years	Self-administered	Legume fiber	Adjusted by age, ethnicity, time since

Reference, cohort, country	Follow-up period	Study size, case no.	Sex, age	Diet assessment	Grain legume, quantity for comparison, risk estimates (95% CI)	Adjustment for confounders
Cohort Study (MEC)				FFQ with 180 food items	Men: Median 7.6 vs. 0.3 g/1000 kcal/d CRC: RR = 0.81 (0.65–1.01) P_{trend} = 0.04 Women: Median 5.8 vs. 0.2 g/1000 kcal/d CRC: RR = 1.02 (0.82–1.27)	cohort entry, and age at cohort
Park et al., 2007 [67] U.S., NIH–AARP Diet and Health Study	1995–2000	488,043, 2972	Both, 50–71 years at baseline	Self-administered FFQ with 124 food items for previous yr	Grain legumes (dried beans, green beans, and peas) Men: Median 0.69 vs. 0.08 servings/d RR = 0.95 (0.83–1.09) Significant for age adjusted RR = 0.85 (0.74–0.97) Women: Median 0.81 vs. 0.09 servings/d RR = 1.13 (0.91–1.40)	Adjusted for education, physical activity, smoking, alcohol consumption, and intake of total energy, red meat, and calcium
Schatzkin et al., 2007 [68] U.S., NIH–AARP Diet and Health Study	1995–2000	489,611, 2974	Both, 50–71 years at baseline	Self-administered FFQ with 124 food items for previous yr	Grain legume fiber Median 2.3 vs. 0.2 g/1000 kcal/d RR = 0.93 (0.83–1.04) Significant for age-and sex-adjusted RR = 0.89 (0.79–0.99)	Adjusted for sex, physical activity, smoking, menopausal hormone therapy, and intake of total energy, red meat, calcium, and folate

*Statistically significant association of legume consumption and colorectal neoplasia.
CRC: colorectal cancer; CRA: colorectal adenoma; FFF: food frequency questionnaire; HR: hazard ratio; IBD: inflammatory bowel disease; N/A: not available; OR: odds ratio; RR: relative risk; 95% CI: 95% confidence interval.
1 serving of legume equals 0.5 cup of cooked legumes (~90 g) [7].

Table 2. Prospective cohort studies of grain legume consumption and colorectal neoplasia.

For the meta-analysis, we had to exclude the CRC risk estimates of two cohorts because the risk estimates did not include 95% CI [65], leaving us with 1,533,527 participants including 12,408 cases. When comparing the highest versus the lowest legume intake group, we observed, as shown in **Figure 2**, a protective effect of grain legume consumption on colorectal neoplasia (RR = 0.89; 95% CI: 0.59–0.88; *P* = 0.001). The protective effect attenuated from incident CRA (RR = 0.72; 95% CI: 0.60–0.87; *P* <0.001) over prevalent CRA (RR = 0.87; 95% CI: 0.75–1.01; *P* = 0.07) to CRC (RR = 0.93; 95% CI: 0.86–1.01; *P* = 0.08). There was little heterogeneity (18.3%) among studies, which was further decreased after stratifying for neoplastic endpoint (**Figure 2**). No significant publication bias was observed (*P* = 0.13). We observed a nonlinear relationship between legume consumption and colorectal neoplasia, as the protective effect of legume consumption for incident CRA (**Table 2**) was limited to the highest legume intake group, which corresponds to daily consumption of at least 0.5 servings of legumes (~45 g/d). In comparison, the 2015 U.S. dietary guidelines recommend three servings/wk (~39 g/d), which is lower than six servings/wk of the 2005 guidelines [7, 76].

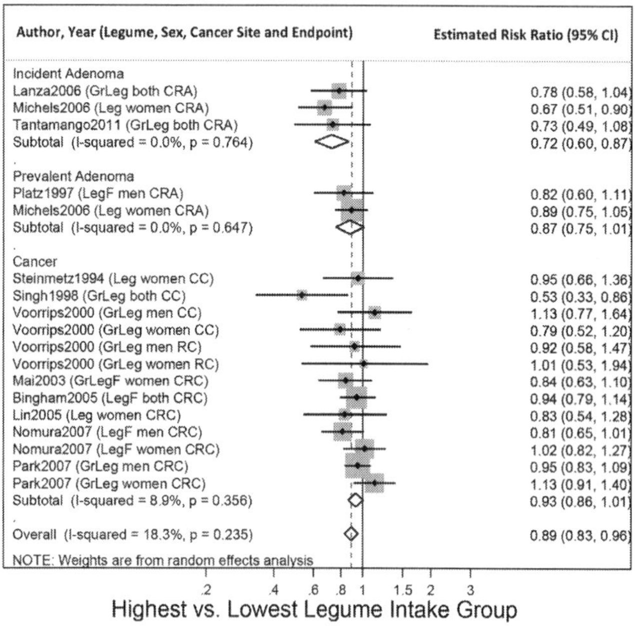

Figure 2. Forest plot of legume consumption (highest vs. lowest category) and colorectal neoplasia risk in prospective studies stratified by type of neoplastic lesion. The dot in each study indicates the estimated risk ratio, vertical bars represent the 95% CI and the size of gray square box reflects the study's weight in the random effects Meta-analysis studies. The straight line indicates no association and the dashed line indicates the summary risk estimate across all studies. The open diamond on the bottom indicates the pooled risk estimate and the right vertices of the diamond reflect the 95% CI. CC: colon cancer; RC: rectal cancer; CRC: colorectal cancer; CRA: colorectal adenoma; GrLeg: grain legume; GrLegF: grain legume fiber; Leg: legume; LegF: legume fiber.

Our risk estimates (**Table 3**) are similar to those obtained previously from meta-analyses between legume consumption (including soybeans) and CRA (RR = 0.73; 95% CI: 0.61–0.88) and CRC (RR = 0.91; 95% CI: 0.84–0.98) [14, 15], as well as legume fiber consumption and CRC (RR = 0.89; 95% CI: 0.78–1.02) [16]. Thus, we conclude that there is limited evidence suggesting that daily grain legume consumption decreases CRC risk in humans, all of which are based on observational studies. This is consistent with what has been previously concluded for the evidence on the relation between stomach or prostate cancer risk and legume consumption [34].

Factor	Studies	Pooled risk ratio		Heterogeneity		Eggers	References
	(estimates)	RR (95% CI)	*P*	I^2 (%)	*P*	*P*	
Overall	23 (36)	0.84 (0.78–0.90)	0.005	41.9	<0.001	0.02	[46–52, 55–57, 59, 60, 62–64, 66, 67, 69–75]
Endpoint							
Incident adenoma	3 (3)	0.72 (0.60–0.87)	<0.001	0	0.76	0.90	[72, 73, 75]
Prevalent adenoma	8 (10)	0.91 (0.84–0.99)	0.03	0	0.73	0.60	[47–52, 73, 74]
Cancer	14 (23)	0.82 (0.74–0.91)	<0.001	54.4	0.001	0.02	[46, 55–57, 59, 60, 62–64, 66, 67, 69–71]
Study type							
Retrospective	12 (18)	0.77 (0.66–0.89)	<0.001	53.3	0.004	0.11	[46–52, 55–57, 59, 60]
Prospective	11 (18)	0.89 (0.83–0.96)	0.001	18.3	0.24	0.13	[62–64, 66, 67, 69–75]
Gender							
Men	10 (11)	0.89 (0.81–0.97)	0.009	0	0.80	0.40	[46, 47, 49, 59, 60, 66, 67, 71, 72, 74]
Women	11 (13)	0.86 (0.75–0.98)	0.03	50.7	0.02	0.14	[46, 47, 49, 59, 60, 64, 66, 67, 70, 71]
Legume type							
Legume	13 (17)	0.88 (0.82–0.94)	<0.001	4.5	0.40	0.10	[48, 49, 51, 57, 59, 60, 62, 63, 66, 70, 73, 74]
Grain legume	11 (19)	0.80 (0.71–0.92)	0.001	58.1	0.001	0.09	[46, 47, 50, 52, 55, 56, 64, 67, 69, 71, 72, 75]
Legume part							
Grain	18 (29)	0.82 (0.74–0.89)	<0.001	49.6	0.001	0.01	[46, 48–52, 55–57, 59, 60, 63, 67, 69–73, 75]
Fiber	6 (8)	0.92 (0.85–0.99)	0.02	0	0.78	0.92	[47, 62, 64, 66, 68, 74]

Factor	Studies	Pooled risk ratio		Heterogeneity		Eggers	References
	(estimates)	RR (95% CI)	P	I^2 (%)	P	P	
Cancer site							
Colon	8 (10)	0.69 (0.54–0.88)	0.003	63.6	0.003	0.94	[45, 55–57, 60, 69–71]
Rectum	3 (4)	0.70 (0.49–1.00)	0.05	63.9	0.04	0.69	[45, 55, 71]
Continent/country							
Europe	4 (8)	0.83 (0.67–1.03)	0.09	64.9	0.006	0.77	[55, 57, 62, 71]
USA	16 (23)	0.88 (0.82–0.94)	<0.001	24.5	0.14	0.04	[47–52, 59, 63, 64, 66, 67, 69, 70, 72–75]

Pooled risk estimates with 95% confidence intervals in parentheses compare risk of developing colorectal cancer/adenoma of the highest versus the lowest grain legumes intake group. Study number will not add up to overall number because for overall study we used most combined risk estimates available. Eggers P-value indicates probability for publication bias.

Table 3. Higher grain legume consumption decreases risk of colorectal tumorigenesis: meta-analysis of 23 human studies.

The next step needs to be a long-term intervention study of daily grain legume consumption in a high CRC risk cohort. Dietary compliance will be a major challenge in Western countries because <10% of the population consumes grain legumes on a daily basis [6, 10, 11]. Moreover, it is much easier to take a daily supplement or a medication than consuming a chemo-preventive diet. At the same time, it is unrealistic to expect a chemo-preventive effect of a food, supplement, or medication when it is sporadically consumed. We previously identified markers of dietary compliance for grain legume consumption in human and animal studies [77], which allows for compliance monitoring. Intention-to-treat analysis, the gold standard for statistical evaluation of intervention studies, assumes high compliance. Statistical methods that account for dietary exposure markers and low compliance are needed when evaluating the evidence from dietary intervention studies.

4. Grain legumes and colorectal neoplasia in animal studies

As shown in **Table 4**, 14 animal studies evaluated the effect of grain legume consumption on colorectal tumorigenesis using 253 animals (248 males and five females) on control diets and 355 animals (350 males and five females) on 19 different grain-legume-containing diets [78–89]. Eight diets contained whole dry beans, seven contained dry bean fractions (three fiber factions, three ethanol extract, and one ethanol extract residue); two diets each contained lentils or chickpeas, and one diet each contained black-eyed peas or dry peas. In three studies, the animals were intragastrically tubed with dry beans and/or dry bean fiber [85, 87], whereas in the remaining 11 studies grain legumes or their fractions were included in the diet. Ten studies were conducted with rats and four with mice. All but one study [79] used azoxymethane

(AOM), which is commonly used in animal models of human CRC to induce DNA mutations by alkylating DNA primarily at the O^6-guanidine residues [90, 91]. After AOM induction, we promoted tumor formation in two unpublished studies with the colon irritant dextran sodium sulfate (DSS); this is an established inflammation-associated animal model of human CRC [92]. Bean treatment started before tumor induction in nine studies, after tumor induction in three, and after tumor induction and promotion in two studies. Study endpoints were ACF in seven studies, adenomas and adenocarcinomas in five, and tumors in two studies.

Reference	Animal	Diet, animals/diet	Experimental design	Tumor endpoints
Colorectal tumors:				
Hughes et al., 1997 [78]	F344 male rats	Control: casein diet, *n* = 20 Treatment: Pinto beans (59% of diet) *n* = 21	2× AOM (15 mg/kg BW) a wk apart First AOM: 6 wk of age Diet Start: 1 wk after last AOM Study End: 34 wk after last AOM	Colon adenomas, adenocarcinomas (incidence and multiplicity)
McIntosh et al., 1998 [79]	Sprague-Dawley male rats	Control: modified AIN-1976, *n* = 18 Treatment: Chickpeas (45% of diet) *n* = 18	3× DMH (15 mg/kg BW) a wk apart First DMH: 9 wk of age Diet start: 4 wk before first DMH Study End: 22 wk after last DMH	Colon adenomas + adenocarcinomas (incidence and multiplicity)
Hangen & Bennink, 2002 [80]	F344 male rats	Control: modified AIN-93G, *n* = 28 Treatments: Black beans (75% of diet) *n* = 32 Navy beans (75% of diet) *n* = 28	2× AOM (15 mg/kg BW) a wk apart First AOM: 7 wk of age Diet Start: 4 wk before first AOM Study End: 31 wk after last AOM	Colon adenomas, adenocarcinomas (incidence and multiplicity)
Bobe et al., 2008 [81]	Ob/Ob male mice	Control: modified AIN-93G, *n* = 40 Treatments: Navy beans (74% of diet) *n* = 34 Navy bean ethanol residue (74% of diet) *n* = 38 Navy bean ethanol extract (9% of diet) n=39	2× AOM (7 mg/kg BW) a wk apart First AOM: 7 wk of age Diet Start: 1 wk after last AOM Study End: 27 wk after last AOM	Colon adenomas, adenocarcinomas, tumors (incidence and multiplicity)

Reference	Animal	Diet, animals/diet	Experimental design	Tumor endpoints
Rondini & Bennink, 2012 [82]	F344 male rats	Control: modified AIN-93G, $n = 25$ Treatment: Black beans (74% of diet) $n = 25$	2× AOM (15 mg/kg BW) a wk apart First AOM: 4 wk of age Diet Start: 1 wk after last AOM Study End: 31 wk after last AOM	Colon adenomas + adenocarcinomas incidence
Bobe et al. (unpublished)	FVB/N male mice	Control: AIN-93G, $n = 32$ Treatment: Navy bean ethanol extract (10% of diet) $n = 33$	AOM (10 mg/kg BW) 6 wk of age DSS (2% drinking water) 1 week starting 1 wk after DSS Diet Start: 10 days after AOM Study End: 102 days after AOM	Colorectal tumor multiplicity
Bobe et al. (unpublished)	FVB/N male mice	Control: AIN-93G, $n = 20$ Treatment: Navy bean ethanol extract (10% of diet) $n = 20$	AOM (10 mg/kg BW) 6 wk of age DSS (2% drinking water) 1 week starting 1 wk after DSS Diet Start: 10 days after AOM Study End: 53 days after AOM	Colorectal tumor multiplicity
Colon aberrant crypt foci (ACF):				
Rijken et al., 1999 [83]	Sprague-Dawley male rats	Control: AIN-93M, $n = 15$ Treatment: Dry peas (5.9% of diet) $n = 15$	2× AOM (15 mg/kg BW) 3 d apart First AOM: 10 wk of age Diet Start: 2 wk before first AOM Study End: 11 wk after last AOM	Colon aberrant crypt foci (total, multiplicity)
Murillo et al., 2004 [84]	CF-1 female mice	Control: Harland Teklad 4% Diet 7001, $n = 5$ Treatment: Chickpea flour (10% of diet) $n = 5$	2× AOM (10 mg/kg BW) a wk apart First AOM: 5 wk of age Diet Start: 2 wk before first AOM Study End: 7 wk after last AOM	Control: 1.13 ACF/cm^2 colon 0 >4 foci ACF Chickpea: 0.41 ACF/cm^2 colon 2.2 ± 0.37 >4 foci ACF
Boateng et al., 2007 [89]	F344 male rats	Control: AIN-93G, $n = 8$ Treatments: Pinto beans (20% of diet) $n = 8$	2× AOM (15 mg/kg BW) a wk apart First AOM: 7 wk of age Diet Start: 3 wk before	Control: 183 ± 23 ACF Pinto: 64 ± 8 ACF Peas: 40 ± 4 ACF

Reference	Animal	Diet, animals/diet	Experimental design	Tumor endpoints
		Black-eyed peas (20% of diet) $n = 8$	first AOM Study End: 9 wk after last AOM	
Feregrino-Perez et al., 2008 [85]	Sprague-Dawley male rats	Control:2018S Harland Teklad $n = 10$ Treatments: Daily intragastric tubing Dry bean Negro 8025 (3.2 g/kg BW) $n = 10$ Dry bean Negro 8025 fiber fraction (1.84 g/kg BW) $n = 10$	2× AOM (15 mg/kg BW) a wk apart First AOM: 5 wk of age Diet Start: 1 wk before first AOM Study End: 5 wk after last AOM	Distal colon zone: Control: 4.2 ± 0.6 ACF Dry bean: 2.2 ± 0.6 ACF Fiber fraction: 2.0 ± 0.8 ACF Using DAPI stain
Faris et al., 2009 [86]	F344 male rats	Control: AIN-93G, $n = 10$ Treatments: Whole lentils (5% of diet) $n = 10$ Split lentils (5% of diet) $n = 9$	2× AOM (15 mg/kg BW) a wk apart First AOM: 10 wk of age Diet Start: 5 wk before first AOM Study End: 17 wk after last AOM	Control: 178 ± 24 ACF 12.0 ± 1.04 >3 foci ACF Dry bean: 70 ± 8 ACF 2.66 ± 0.09 >3 foci ACF Fiber fraction: 94 ± 17 ACF 5.56 ± 1.05 >3 foci ACF
Vergara-Castaneda et al., 2010 [87]	Sprague-Dawley rats male	Control:2018S Harland Teklad $n = 12$ Treatments: Daily intragastric tubing Dry bean Bayo Madero (5.7 g/kg BW) $n = 12$ Dry bean Bayo Madero fiber fraction (2.5 g/kg BW) $n = 10$	2× AOM (15 mg/kg BW) a wk apart First AOM: 6 wk of age Diet Start: 1 wk before first AOM Study End: 7 wk after last AOM	Distal colon zone: Control: 6.6 ± 0.40 ACF Dry bean: 0.8 ± 0.20ACF Fiber fraction: 1.5 ± 0.72 ACF
Feregrino-Perez et al., 2014 [88]	Sprague-Dawley male rats	Control:2018S Harland Teklad $n = 10$ Treatments: Daily intragastric tubing Dry bean Negro 8025 fiber fraction (1.84 g/kg BW) $n = 10$	2× AOM (15 mg/kg BW) a wk apart First AOM: 5 wk of age Diet Start: 1 wk before first AOM Study End: 5 wk after last AOM	Distal colon zone: Control: 21.0 ± 3.25 ACF Fiber fraction: 7.20 ± 2.95 ACF

AOM: azoxymethane; BW: body weight; DMH: dimethylhydrazine; DSS: dextran sodium sulfate. ACF were measured using methylene blue staining unless otherwise noted.

Table 4. Experimental design and endpoints in animal studies of grain legume intake and colorectal tumorigenesis.

Table 5 shows individual and pooled risk estimates of the seven studies with tumor endpoints. For calculating risk estimates of tumor and ACF multiplicity, we calculated standardized mean differences and variation from reported means and standard errors. Grain legume consumption inhibited colorectal tumorigenesis. The protective effect of dry bean consumption attenuated with progressive tumor stage from tumor incidence (OR = 0.21; 95% CI: 0.11–0.43) over combined adenoma and adenocarcinoma incidence (OR = 0.32; 95% CI: 0.17–0.60) to adenocarcinoma incidence (OR = 0.38; 95% CI: 0.20–0.74). Similarly, the protective effect of grain legume consumption attenuated from ACF multiplicity (OR = 0.07; 95% CI: 0.03–0.14 with stronger effect on larger ACFs; **Table 4**) over tumor multiplicity (OR = 0.24; 95% CI: 0.16–0.36) to combined adenoma and adenocarcinoma multiplicity (OR = 0.52; 95% CI: 0.31–0.89) and adenocarcinoma multiplicity (OR = 0.52; 95% CI: 0.27–0.98; *P* = 0.04). Given that the chemopreventive effect of legumes was reported when grain legumes were fed before as well as after tumor induction and/or tumor promotion, we conclude that grain legumes inhibit colorectal tumorigenesis at different tumor stages.

Reference, Year	**Legume**	**Adenocarcinoma**		**Adenoma + adenocarcinoma**		**Tumor**	
		Incidence	**Multiplicity**	**Incidence**	**Multiplicity**	**Incidence**	**Multiplicity**
Hughes1997	PintoBW	0.38 (0.10–1.45)	0.19 (0.06–0.60)	0.31 (0.08–1.19)	0.20 (0.06–0.66)		
Hangen2002	BlackBW	0.19 (0.05–0.77)		0.25 (0.09–0.75)			
Bennink2012	BlackBW			0.15 (0.04–0.52)			
Hangen2002	NavyBW	0.30 (0.08–1.11)		0.22 (0.07–0.68)			
Bobe2008	NavyBW	1.55 (0.38–6.31)	1.11 (0.48–2.55)	0.59 (0.18–1.98)	0.90 (0.39–2.07)	0.32 (0.11–0.95)	0.29 (0.12–0.68)
Bobe2008	NavyBER	0.24 (0.03–2.28)	0.56 (0.25–1.26)	0.23 (0.07–0.71)	0.61 (0.27–1.36)	0.23 (0.07–0.71)	0.22 (0.09–0.51)
Bobe2008	NavyBEE	0.23 (0.02–2.16)	0.45 (0.20–1.01)	0.09 (0.01–0.74)	0.46 (0.21–1.04)	0.08 (0.02–0.38)	0.17 (0.07–0.39)
BobeUnpubl	NavyBEE						0.20 (0.05–0.74)
BobeUnpubl	NavyBEE						0.34 (0.14–0.85)
McIntosh1998	ChickpeaW			2.50 (0.65–9.65)			
Pooled odds ratio		**0.38 (0.20–0.74)**	**0.52 (0.27–0.98)**	**0.32 (0.17–0.60)**	**0.52 (0.31–0.89)**	**0.21 (0.11–0.43)**	**0.24 (0.16–0.36)**

For multiplicity, odds ratios and their 95% confidence intervals were estimated from reported means and standard errors by calculating standardized mean differences. B: bean; BEE: bean ethanol extract; BER: bean ethanol residue; W: whole beans; multiplicity: number of tumor/animal.
The P-values are in this order from left to right: P=0.004; P = 0.04; P< 0.001; P = 0.02; P< 0.001; P< 0.001

Table 5. Risk estimates with 95% confidence intervals (in parentheses) for colorectal tumors in animal studies.

The animal studies have limitations: first, in four of the seven tumor endpoint studies, grain legumes made up the majority of the diet (45–75%; **Table 4**) [78–80, 82], concentrations that are not relevant for human consumption. However, three studies showed a protective effect of the ethanol extract of navy beans fed at 10% of the diet (**Table 4**); the 2015 U.S. dietary guidelines for legume consumption are equivalent to ~2–5% of the diet [76], concentrations that should be evaluated in future animal studies. Second, none of the reported studies included more than one grain legume dosage (**Table 4**), demonstrating a need for dose-response studies in animal CRC models. Third, only one study examined the chemo-preventive effect of grain legumes other than dry beans at the tumor stage (**Table 4**), indicating a need to evaluate the chemo-preventive effect of other grain legumes at the tumor stage. Fourth, further research is needed to demonstrate a chemo-preventive response in female animals, as all but one study [84] examined the response in male animals. Despite these limitations, there is sufficient evidence to conclude that at least dry bean consumption probably decreases colorectal tumorigenesis in male animal models of human CRC.

5. Chemo-preventive compounds in grain legumes

To elucidate which fractions of grain legumes have chemo-preventive properties against colorectal tumorigenesis, we previously fractionated cooked navy beans using 60% ethanol [81]. Both the ethanol extract and the residue inhibited colorectal tumorigenesis in AOM-induced mice, indicating that both fractions contain chemo-preventive compounds. Several studies conducted by Loarca-Piňa's research group demonstrated that the non-digestible fraction of dry beans inhibits colon ACF formation in AOM-induced rats [85, 87].

Grain legumes contain three major carbohydrate classes that inhibited colorectal ACF and tumor formation in animal CRC models: resistant starches (cooked grain legumes contain 0.6–4.2%), soluble fiber including the flatulence-inducing α-galacto-oligosaccharides stachyose, verbascose, and raffinose (cooked grain legumes contain 0–3%), and insoluble fiber (cooked grain legumes contain 15–23%); concentrations of those carbohydrate classes vary considerably based on processing methods [1, 2, 7, 93–97]. Resistant starches can be effective at 5–10% of the diet [7, 98–102]. Soluble fiber can inhibit ACF and tumor formation at 2.5–15% of the diet [103, 104], and insoluble fiber can be effective at 5–15% of the diet [104–107].

Grain legumes contain lipid classes that inhibited colorectal ACF and tumor formation in animal models of CRC. Plant sterols (e.g., β-sitosterol, campesterol, and stigmasterol; 0.13–0.24% of grain legume dry weight) attenuate colorectal tumorigenesis in animal studies (gastric intubation of 10–20 mg β-sitosterol/kg body weight or 0.2% of diet) [108–111]. Saponins (0.1–0.5% of grain legume dry weight) are glycolipids, which inhibit ACF formation at concentrations of 0.01–3% of the diets [112–116]; the lower concentrations are relevant for human diets [117]. Processing can decrease saponin concentrations in grain legumes up to 40% [118]. Besides containing phytosterols and saponins, grain legumes are low in lipids and have a favorable fatty acid composition for chemo-prevention (i.e., low in saturated fatty acids and a low $\Omega3$: $\Omega6$ fatty acid ratio) [3, 119, 120].

Grain legumes contain protein classes that inhibited colorectal ACF and tumor formation in animal models of CRC. Trypsin and chymotrypsin protease inhibitors of the Bowman-Birk family inhibit at dietary concentrations of 0.1–0.5% of the diet or 20 mg/kg of body weight for colorectal ACF and tumor formation [121–125]. Lectins (i.e., agglutinins; 0.1–3.5% of grain legume dry weight), which are glycoproteins that bind to epithelial cells, have been shown to inhibit cancer growth in animal tumor transplant studies and colon cancer cells [126–128]. Grain legumes have significant α-amylase inhibitor activity, which may indirectly decrease CRC risk by increasing microbial butyrate production and decreasing blood glucose and insulin after starch consumption [129]. The importance of Bowman-Birk inhibitors, α-amylase inhibitors, and lectins is debatable because 80–90% is lost and denatured during soaking and cooking, respectively [7, 96, 117].

The mineral and vitamin content of grain legumes may confer chemo-preventive effects against colorectal tumorigenesis. Grain legumes contain high concentrations of folate (83–174 μg/100 g of cooked legumes) and potassium (0.29–0.51% of cooked legumes) and low concentrations of sodium (<0.01% of cooked legumes) [7]. A high ratio of potassium to sodium has been reported to decrease CRC risk, and folate intake is established as a protective nutrient against CRC [130, 131]. Chemo-preventive compounds associated with minerals are phytates (0.1–1.9% of grain legume dry weight), the primary plant storage forms of phosphorus [117]. Processing decreases phytate content up to 50% [97, 132]. Phytates inhibit ACF formation at dietary concentrations of 0.02–2% [133–136]; the lower concentrations are relevant for human diets [137].

Grain legumes are a good dietary source of phenolic compounds (1–10 mg gallic acid equivalents/g legume, which is ~0.1–1.0% of grain legume dry weight) [117, 118, 132, 138, 139], many of which inhibited colorectal ACF and tumor formation in animal models of CRC. The three major phenolic groups with chemo-preventive properties are flavonoids (0–5 mg catechin equivalents/g legume), proanthocyanidins (i.e., condensed tannins; 0.2–12 mg catechin equivalents/g legume), and phenolic acids (0.02–0.1% of cooked legume dry weight) [118, 132, 138, 139]. Flavonols (i.e., kaempferol and quercetin), anthocyanidins, and flavan-3-ols are major flavonoid classes in grain legumes that have been demonstrated by us and others to inhibit colorectal tumor multiplicity at concentrations of 0.05–0.3% of the diet [140–144]. Proanthocyanidins can inhibit ACF formation at concentrations of 0.002–1% of the diet or by gavage [145–147]. Phenolic acids include ferulic acid (~0.003% of grain legume dry weight) that inhibited ACF formation at concentrations of 0.25–1% [148–150] and sinapic acid that inhibited ACF formation at concentrations of 20–80 mg/kg of body weight by gavage [151]. The concentrations of the phyto-estrogen group's isoflavonoids (0.005–0.095 mg/kg grain legume) and lignans (0.018–0.266 mg/kg grain legume) are relatively low in grain legumes [152] and, thus, probably contribute little to the chemo-preventive effect of grain legumes. Processing and cooking of grain legumes result in various losses of phenolic compounds, which decreased not only their antioxidant activities but also their antiproliferative properties against colon cancer cells [118, 132, 139]. Thus, food processing plays an important role for the chemo-preventive role of grain legumes [117, 127].

There is sufficient evidence that grain legumes contain various compounds that can exert chemo-preventive effects against colorectal tumorigenesis in animal models of CRC at concentrations that are relevant for human diets. One has to consider that several of the aforementioned compounds are developed by plants as defense mechanisms against herbivores and are at sufficiently high concentrations to be toxic. It has to be noted that most of the aforementioned compounds do not show a consistent chemo-preventive effect in animal models of CRC; further investigation is necessary to elucidate factors, including food processing, that affect the response. Further studies are also warranted to examine whether the effect of the chemo-preventive compounds differs when they are consumed alone or in combination.

6. Molecular mechanisms by which grain legumes inhibit colorectal tumorigenesis

Given the complex mixture of chemo-preventive compounds in grain legumes, it comes to no surprise that grain legumes inhibit hallmarks of cancer [153, 154] at multiple stages of the colorectal tumorigenesis process. (A) Grain legumes can inhibit tumor induction (i.e., the transition from normal to initiated colorectal epithelial cells). First, grain legumes can alter the metabolism of carcinogens (i.e., increased degradation) and pre-carcinogens (i.e., decreased activation). This is accomplished directly by activating the expression of cytochrome P450 and UDP-glucuronosyltransferase (UGT) protein-encoding genes in the liver and indirectly by altering microbiome metabolism of carcinogens (e.g., decreased β-glucuronidase activity) in the colon [87, 155]. Second, grain legumes can act as antioxidants and induce genes involved in the detection and repair of mutated genes [156, 157]. Third, grain legumes may prevent the exposure of colorectal epithelial cells to carcinogens in food and bile by (a) binding carcinogens with non-digestible grain legume compounds [87, 158] and by (b) increasing mucin production of colorectal epithelial cells [159]. Fourth, grain legumes can decrease the colon pH [80] and promote the growth of probiotic bacteria [160] and thereby inhibit the growth of genotoxic bacteria [161, 162].

(B) Grain legumes can inhibit tumor promotion and progression (i.e., the transformation from initiated to neoplastic colorectal epithelial cells). First, grain legumes can increase apoptosis through the mitochondrial-mediated and death receptor-mediated pathways in neoplastic colorectal epithelial cells [88, 156] and colon cancer cell lines [163–165]. Second, grain legumes can inhibit survival of neoplastic colorectal epithelial cells by attenuating the NF-kB pathway [163–165]. Third, grain legumes can decrease proliferation of neoplastic colorectal epithelial cells [156, 163] by inducing genes that promote cell cycle arrest in G1/S and G2/M phases through p53-mediated pathways [82, 156, 165]. Fourth, grain legumes can inhibit survival and proliferation of neoplastic cells by suppressing the Akt (protein kinase B)/mTOR (mammalian target of rapamycin) pathway and upregulating the AMPK pathway, as shown for mammary carcinomas [166, 167]. In addition, upregulation of the AMPK and p53 pathway and suppression of the Akt/mTOR pathway may limit the nutrient and energy supply for the rapidly growing cancer cells and thereby inhibit tumor growth and progression [168–170]. Fifth, grain legumes can inhibit survival and proliferation of neoplastic colorectal epithelial cells through increased butyrate production in the colon [80, 163, 171].

(C) Grain legumes can inhibit tumor promotion and progression indirectly by limiting and/or resolving inflammation. Inflammation creates a tumor microenvironment that encourages neoplastic transformations and promotes survival and proliferation of neoplastic colorectal epithelial cells. We previously showed in the Polyp Prevention Trial that the chemo-preventive effect of grain legumes against CRA recurrence is linked to a decrease in serum interleukin (IL)-6 [172]. Moreover, we demonstrated in AOM-induced ob/ob mice that navy beans and their ethanol extract decreased concomitantly colorectal neoplasia and IL-6 in serum and colon mucosa [173]. In support, others demonstrated that grain legumes can attenuate the DSS-induced increase in serum cytokine concentrations [139, 159]. Multiple mechanisms are involved in the anti-inflammatory effect of grain legumes: first, grain legume fractions can act as antioxidants and inhibit NF-kB pathways and gene expression of COX-2 and tumor necrosis factor (TNF)-α [165, 174]; second, grain legume consumption can increase mucin gene expression in the colon and thereby preserve epithelial integrity during inflammation [82, 159]; third, grain legumes can promote microbial butyrate production in the colon, which has anti-inflammatory and antitumor effects [175]; fourth, grain legumes can promote the growth of probiotic bacteria [160] and thereby inhibit the growth of inflammation-inducing bacteria [162, 176].

There is sufficient evidence in human studies, animal models, and colon cancer cell lines for multiple molecular pathways/mechanisms by which grain legume consumption inhibits early stages of colorectal tumorigenesis (i.e., tumor induction, promotion, and progression). The main molecular mechanisms involved are preventing genotoxic hits, DNA repair, inhibiting survival and proliferation of neoplastic colorectal epithelial cells, preventing, limiting, and/or resolving inflammation, and limiting nutrient supply for neoplastic colorectal epithelial cells. Identification of grain legume response biomarkers (i.e., indicators that are linked to both grain legume consumption and inhibition of colorectal tumorigenesis such as IL-6) will be important to evaluate the efficacy of grain legumes in future long-term intervention studies in humans. Grain legume consumption alters the composition and metabolism of colon microbiota, cell cycle kinetics, and metabolism of colorectal epithelial cells, as well as host immune response and barrier function of the colon. Future studies are warranted to examine how grain legumes and their components alter the interplay between microbiota and host. Furthermore, more research is needed to understand the effect of grain legumes on the later stages of colorectal carcinogenesis (i.e., metastasis and invasion).

7. Conclusions

The objective of this chapter was to evaluate the evidence of a chemo-preventive role of grain legume consumption in colorectal tumorigenesis. Based on a literature review and meta-analyses, we conclude that there is limited evidence from case-control and cohort studies suggesting that daily grain legume consumption decreases CRC risk in humans. There is considerable preclinical evidence in animal models of CRC that supports a chemo-preventive effect of dry beans in male animal CRC models. There is sufficient evidence that grain legumes contain various compounds that can exert chemo-preventive effects against colorectal tumorigenesis in animal models of CRC. This is accomplished at concentrations that are relevant for human diets through multiple molecular pathways, which are critical for induction and clonal expansion of neoplastic colorectal epithelial cells. In summary, on the basis of the current evidence, daily grain legume consumption confers chemo-preventive effects against CRC. The next step is to conduct a long-term grain legume CRC prevention intervention study in humans to further elucidate the effects of daily grain legume consumption using grain legume exposure biomarkers to validate compliance and grain legume response biomarkers to monitor efficacy.

Author details

Thushanthi Perera, Yumie Takata and Gerd Bobe*

*Address all correspondence to: Gerd.Bobe@oregonstate.edu

College of Public Health and Human Sciences, Linus Pauling Institute, Linus Pauling Science Center, Oregon State University, Corvallis, OR, USA

References

[1] Mudryj AN, Yu N, Aukema HM. Nutritional and health benefits of pulses. Applied physiology, nutrition, and metabolism. Appl Physiol Nutr Metab. 2014;39:1197–204.

[2] Rebello CJ, Greenway FL, Finley JW. Whole grains and pulses: a comparison of the nutritional and health benefits. J Agric Food Chem. 2014;62:7029–49.

[3] Geil PB, Anderson JW. Nutrition and health implications of dry beans - a review. J Am Coll Nutr. 1994;13:549–58.

[4] Sanchez-Chino X, Jimenez-Martinez C, Davila-Ortiz G, Alvarez-Gonzalez I, Madrigal-Bujaidar E. Nutrient and nonnutrient components of legumes, and its chemopreventive activity: a review. Nutr Cancer. 2015;67:401–10.

[5] Duranti M. Grain legume proteins and nutraceutical properties. Fitoterapia. 2006;77:67–82.

[6] Mitchell DC, Lawrence FR, Hartman TJ, Curran JM. Consumption of dry beans, peas, and lentils could improve diet quality in the US population. J Am Diet Assoc. 2009;109:909–13.

[7] Messina V. Nutritional and health benefits of dried beans. Am J Clin Nutr. 2014;100 Suppl 1:437S–42S.

[8] Roy F, Boye JI, Simpson BK. Bioactive proteins and peptides in pulse crops: pea, chickpea and lentil. Food Res Int. 2010;43:432–42.

[9] Bejar LM, Gili M, Infantes B, Marcott PF. Effects of changes in dietary habits on colorectal cancer incidence in twenty countries from four continents during the period 1971–2002. Rev Esp Enferm Dig. 2011;103:519–29.

[10] Mudryj AN, Yu N, Hartman TJ, Mitchell DC, Lawrence FR, Aukema HM. Pulse consumption in Canadian adults influences nutrient intakes. Br J Nutr. 2012;108 Suppl 1:S27–36.

[11] Schneider AV. Overview of the market and consumption of pulses in Europe. Br J Nutr. 2002;88 Suppl 3:S243–50.

[12] Guenther PM, Dodd KW, Reedy J, Krebs-Smith SM. Most Americans eat much less than recommended amounts of fruits and vegetables. J Am Diet Assoc. 2006;106:1371–9.

[13] Bray F, Ren JS, Masuyer E, Ferlay J. Global estimates of cancer prevalence for 27 sites in the adult population in 2008. Int J Cancer. 2013;132:1133–45.

[14] Wang Y, Wang Z, Fu L, Chen Y, Fang J. Legume consumption and colorectal adenoma risk: a meta-analysis of observational studies. PLoS One. 2013;8:e67335.

[15] Zhu B, Sun Y, Qi L, Zhong R, Miao X. Dietary legume consumption reduces risk of colorectal cancer: evidence from a meta-analysis of cohort studies. Sci Rep. 2015;5:8797.

[16] Aune D, Chan DS, Lau R, Vieira R, Greenwood DC, Kampman E, et al. Dietary fibre, whole grains, and risk of colorectal cancer: systematic review and dose-response meta-analysis of prospective studies. BMJ. 2011;343:d6617.

[17] Ferlay J, Soerjomataram I, Dikshit R, Eser S, Mathers C, Rebelo M, et al. Cancer incidence and mortality worldwide: sources, methods and major patterns in GLOBOCAN 2012. Int J Cancer. 2015;136:E359–86.

[18] Mariotto AB, Yabroff KR, Shao Y, Feuer EJ, Brown ML. Projections of the cost of cancer care in the United States: 2010–2020. J Natl Cancer Inst. 2011;103:117–28.

[19] American Cancer Society. Colorectal Cancer Facts & Figures 2014–2016. Atlanta, GA: American Cancer Society; 2014.

[20] Tomasetti C, Vogelstein B. Variation in cancer risk among tissues can be explained by the number of stem cell divisions. Science. 2015;347:78–81.

[21] Vogelstein B, Papadopoulos N, Velculescu VE, Zhou SB, Diaz LA, Kinzler KW. Cancer genome landscapes. Science. 2013;339:1546–58.

[22] Takayama T, Katsuki S, Takahashi Y, Ohi M, Nojiri S, Sakamaki S, et al. Aberrant crypt foci of the colon as precursors of adenoma and cancer. N Engl J Med. 1998;339:1277–84.

[23] Kinzler KW, Vogelstein B. Lessons from hereditary colorectal cancer. Cell. 1996;87:159–70.

[24] Peipins LA, Sandler RS. Epidemiology of colorectal adenomas. Epidemiol Rev. 1994;16:273–97.

[25] Bujanda L, Cosme A, Gil I, Arenas-Mirave JI. Malignant colorectal polyps. World J Gastroenterol. 2010;16:3103–11.

[26] DeSantis CE, Lin CC, Mariotto AB, Siegel RL, Stein KD, Kramer JL, et al. Cancer treatment and survivorship statistics, 2014. CA Cancer J Clin. 2014;64:252–71.

[27] Siegel R, DeSantis C, Virgo K, Stein K, Mariotto A, Smith T, et al. Cancer treatment and survivorship statistics, 2012. CA Cancer J Clin. 2012;62:220–41.

[28] Lutgens MW, van Oijen MG, van der Heijden GJ, Vleggaar FP, Siersema PD, Oldenburg B. Declining risk of colorectal cancer in inflammatory bowel disease: an updated meta-analysis of population-based cohort studies. Inflamm Bowel Dis. 2013;19:789–99.

[29] Deng L, Gui Z, Zhao L, Wang J, Shen L. Diabetes mellitus and the incidence of colorectal cancer: an updated systematic review and meta-analysis. Dig Dis Sci. 2012;57:1576–85.

[30] Moghaddam AA, Woodward M, Huxley R. Obesity and risk of colorectal cancer: a meta-analysis of 31 studies with 70,000 events. Cancer Epidemiol Biomarkers Prev. 2007;16:2533–47.

[31] Shi J, Xiong L, Li J, Cao H, Jiang W, Liu B, et al. A linear dose-response relationship between fasting plasma glucose and colorectal cancer risk: systematic review and meta-analysis. Sci Rep. 2015;5:17591.

[32] Jiang Y, Ben Q, Shen H, Lu W, Zhang Y, Zhu J. Diabetes mellitus and incidence and mortality of colorectal cancer: a systematic review and meta-analysis of cohort studies. Eur J Epidemiol. 2011;26:863–76.

[33] Giovannucci E. Metabolic syndrome, hyperinsulinemia, and colon cancer: a review. Am J Clin Nutr. 2007;86:s836–42.

[34] World Cancer Research Fund and American Institute for Cancer Research. Food, Nutrition, Physical Activity, and the Prevention of Cancer: a Global Perspective. 2nd ed. Washington, DC: American Institute for Cancer Research; 2007.

[35] Chan AT, Giovannucci EL. Primary prevention of colorectal cancer. Gastroenterology. 2010;138:2029–43.

[36] Pericleous M, Mandair D, Caplin ME. Diet and supplements and their impact on colorectal cancer. J Gastrointest Oncol. 2013;4:409–23.

[37] Bejar L, Gili M, Diaz V, Ramirez G, Lopez J, Cabanillas JL, et al. Incidence and mortality by colorectal cancer in Spain during 1951–2006 and its relationship with behavioural factors. Eur J Cancer Prev. 2009;18:436–44.

[38] Bejar LM, Gili M, Infantes B, Marcott PF. Incidence of colorectal cancer and influence of dietary habits in fifteen European countries from 1971 to 2002. Gac Sanit. 2012;26:69–73.

[39] Bejar Prado LM, Gili M, Ramirez G, Lopez J, Cabanillas JL. Dietary changes and colorectal cancer trends in Spain during 1951–2007. Rev Esp Enferm Dig. 2010;102:159–68.

[40] Correa P. Epidemiological correlations between diet and cancer frequency. Cancer Res. 1981;41:3685–90.

[41] Monroe KR, Hankin JH, Pike MC, Henderson BE, Stram DO, Park S, et al. Correlation of dietary intake and colorectal cancer incidence among Mexican-American migrants: the Multiethnic Cohort Study. Nutr Cancer. 2003;45:133–47.

[42] Haenszel W, Berg JW, Segi M, Kurihara M, Locke FB. Large-bowel cancer in Hawaiian Japanese. J Natl Cancer Inst. 1973;51:1765–79.

[43] Abu Mweis SS, Tayyem RF, Shehadah I, Bawadi HA, Agraib LM, Bani-Hani KE, et al. Food groups and the risk of colorectal cancer: results from a Jordanian case-control study. Eur J Cancer Prev. 2015;24:313–20.

[44] Tayyem RF, Bawadi HA, Shehadah I, Agraib LM, Al-Awwad NJ, Heath DD, et al. Consumption of whole grains, refined cereals, and legumes and its association with colorectal cancer among Jordanians. Integr Cancer Ther. 2015;Dec 1, pii: 1534735415620010 [ahead of print].

[45] Aune D, De Stefani E, Ronco A, Boffetta P, Deneo-Pellegrini H, Acosta G, et al. Legume intake and the risk of cancer: a multisite case-control study in Uruguay. Cancer Causes Control. 2009;20:1605–15.

[46] Deneo-Pellegrini H, Boffetta P, De Stefani E, Ronco A, Brennan P, Mendilaharsu M. Plant foods and differences between colon and rectal cancers. Eur J Cancer Prev. 2002;11:369–75.

[47] Sandler RS, Lyles CM, Peipins LA, McAuliffe CA, Woosley JT, Kupper LL. Diet and risk of colorectal adenomas: macronutrients, cholesterol, and fiber. J Natl Cancer Inst. 1993;85:884–91.

[48] Witte JS, Longnecker MP, Bird CL, Lee ER, Frankl HD, Haile RW. Relation of vegetable, fruit, and grain consumption to colorectal adenomatous polyps. Am J Epidemiol. 1996;144:1015–25.

[49] Smith-Warner SA, Elmer PJ, Fosdick L, Randall B, Bostick RM, Grandits G, et al. Fruits, vegetables, and adenomatous polyps: the Minnesota Cancer Prevention Research Unit case-control study. Am J Epidemiol. 2002;155:1104–13.

[50] Agurs-Collins T, Smoot D, Afful J, Makambi K, Adams-Campbell LL. Legume intake and reduced colorectal adenoma risk in African-Americans. J Natl Black Nurses Assoc. 2006;17:6–12.

[51] Millen AE, Subar AF, Graubard BI, Peters U, Hayes RB, Weissfeld JL, et al. Fruit and vegetable intake and prevalence of colorectal adenoma in a cancer screening trial. Am J Clin Nutr. 2007;86:1754–64.

[52] Wu H, Dai Q, Shrubsole MJ, Ness RM, Schlundt D, Smalley WE, et al. Fruit and vegetable intakes are associated with lower risk of colorectal adenomas. J Nutr. 2009;139:340–4.

[53] Benito E, Stiggelbout A, Bosch FX, Obrador A, Kaldor J, Mulet M, et al. Nutritional factors in colorectal cancer risk: a case-control study in Majorca. Int J Cancer. 1991;49:161–7.

[54] Bidoli E, Franceschi S, Talamini R, Barra S, La Vecchia C. Food consumption and cancer of the colon and rectum in north-eastern Italy. Int J Cancer. 1992;50:223–9.

[55] Franceschi S, Parpinel M, La Vecchia C, Favero A, Talamini R, Negri E. Role of different types of vegetables and fruit in the prevention of cancer of the colon, rectum, and breast. Epidemiology. 1998;9:338–41.

[56] Iscovich JM, L'Abbe KA, Castelleto R, Calzona A, Bernedo A, Chopita NA, et al. Colon cancer in Argentina. I: Risk from intake of dietary items. Int J Cancer. 1992;51:851–7.

[57] Kampman E, Verhoeven D, Sloots L, van't Veer P. Vegetable and animal products as determinants of colon cancer risk in Dutch men and women. Cancer Causes Control. 1995;6:225–34.

[58] La Vecchia C, Negri E, Decarli A, D'Avanzo B, Gallotti L, Gentile A, et al. A case-control study of diet and colo-rectal cancer in northern Italy. Int J Cancer. 1988;41:492–8.

[59] Le Marchand L, Hankin JH, Wilkens LR, Kolonel LN, Englyst HN, Lyu LC. Dietary fiber and colorectal cancer risk. Epidemiology. 1997;8:658–65.

[60] Steinmetz KA, Potter JD. Food-group consumption and colon cancer in the Adelaide case-control study. I. Vegetables and fruit. Int J Cancer. 1993;53:711–9.

[61] Bingham SA, Day NE, Luben R, Ferrari P, Slimani N, Norat T, et al. Dietary fibre in food and protection against colorectal cancer in the European Prospective Investigation into Cancer and Nutrition (EPIC): an observational study. Lancet. 2003;361:1496–501.

[62] Bingham SA, Norat T, Moskal A, Ferrari P, Slimani N, Clavel-Chapelon F, et al. Is the association with fiber from foods in colorectal cancer confounded by folate intake? Cancer Epidemiol Biomarkers Prev. 2005;14:1552–6.

[63] Lin J, Zhang SM, Cook NR, Rexrode KM, Liu S, Manson JE, et al. Dietary intakes of fruit, vegetables, and fiber, and risk of colorectal cancer in a prospective cohort of women (United States). Cancer Causes Control. 2005;16:225–33.

[64] Mai V, Flood A, Peters U, Lacey JV, Jr., Schairer C, Schatzkin A. Dietary fibre and risk of colorectal cancer in the Breast Cancer Detection Demonstration Project (BCDDP) follow-up cohort. Int J Epidemiol. 2003;32:234–9.

[65] Michels KB, Edward G, Joshipura KJ, Rosner BA, Stampfer MJ, Fuchs CS, et al. Prospective study of fruit and vegetable consumption and incidence of colon and rectal cancers. J Natl Cancer Inst. 2000;92:1740–52.

[66] Nomura AM, Hankin JH, Henderson BE, Wilkens LR, Murphy SP, Pike MC, et al. Dietary fiber and colorectal cancer risk: the Multiethnic Cohort Study. Cancer Causes Control. 2007;18:753–64.

[67] Park Y, Subar AF, Kipnis V, Thompson FE, Mouw T, Hollenbeck A, et al. Fruit and vegetable intakes and risk of colorectal cancer in the NIH-AARP Diet and Health Study. Am J Epidemiol. 2007;166:170–80.

[68] Schatzkin A, Mouw T, Park Y, Subar AF, Kipnis V, Hollenbeck A, et al. Dietary fiber and whole-grain consumption in relation to colorectal cancer in the NIH-AARP Diet and Health Study. Am J Clin Nutr. 2007;85:1353–60.

[69] Singh PN, Fraser GE. Dietary risk factors for colon cancer in a low-risk population. Am J Epidemiol. 1998;148:761–74.

[70] Steinmetz KA, Kushi LH, Bostick RM, Folsom AR, Potter JD. Vegetables, fruit, and colon cancer in the Iowa Women's Health Study. Am J Epidemiol. 1994;139:1–15.

[71] Voorrips LE, Goldbohm RA, van Poppel G, Sturmans F, Hermus RJ, van den Brandt PA. Vegetable and fruit consumption and risks of colon and rectal cancer in a prospective cohort study: The Netherlands Cohort Study on Diet and Cancer. Am J Epidemiol. 2000;152:1081–92.

[72] Lanza E, Hartman TJ, Albert PS, Shields R, Slattery M, Caan B, et al. High dry bean intake and reduced risk of advanced colorectal adenoma recurrence among participants in the polyp prevention trial. J Nutr. 2006;136:1896–903.

[73] Michels KB, Giovannucci E, Chan AT, Singhania R, Fuchs CS, Willett WC. Fruit and vegetable consumption and colorectal adenomas in the Nurses' Health Study. Cancer Res. 2006;66:3942–53.

[74] Platz EA, Giovannucci E, Rimm EB, Rockett HR, Stampfer MJ, Colditz GA, et al. Dietary fiber and distal colorectal adenoma in men. Cancer Epidemiol Biomarkers Prev. 1997;6:661–70.

[75] Tantamango YM, Knutsen SF, Beeson WL, Fraser G, Sabate J. Foods and food groups associated with the incidence of colorectal polyps: the Adventist Health Study. Nutr Cancer. 2011;63:565–72.

[76] U.S. Department of Agriculture and U.S. Department of Health and Human Services. Dietary Guidelines for Americans. 8th ed. Washington, DC: Government Printing Office; 2015.

[77] Perera T, Young MR, Zhang ZY, Murphy G, Colburn NH, Lanza E, et al. Identification and monitoring of metabolite markers of dry bean consumption in parallel human and mouse studies. Mol Nutr Food Res. 2015;59:795–806.

[78] Hughes JS, Ganthavorn C, Wilson-Sanders S. Dry beans inhibit azoxymethane-induced colon carcinogenesis in F344 rats. J Nutr. 1997;127:2328–33.

[79] McIntosh GH, Wang YH, Royle PJ. A diet containing chickpeas and wheat offers less protection against colon tumors than a casein and wheat diet in dimethylhydrazine-treated rats. J Nutr. 1998;128:804–9.

[80] Hangen L, Bennink MR. Consumption of black beans and navy beans (*Phaseolus vulgaris*) reduced azoxymethane-induced colon cancer in rats. Nutr Cancer. 2002;44:60–5.

[81] Bobe G, Barrett KG, Mentor-Marcel RA, Saffiotti U, Young MR, Colburn NH, et al. Dietary cooked navy beans and their fractions attenuate colon carcinogenesis in azoxymethane-induced ob/ob mice. Nutr Cancer. 2008;60:373–81.

[82] Rondini EA, Bennink MR. Microarray analyses of genes differentially expressed by diet (black beans and soy flour) during azoxymethane-induced colon carcinogenesis in rats. J Nutr Metab. 2012;2012:351796.

[83] Rijken PJ, Timmer WG, van de Kooij AJ, van Benschop IM, Wiseman SA, Meijers M, et al. Effect of vegetable and carotenoid consumption on aberrant crypt multiplicity, a surrogate end-point marker for colorectal cancer in azoxymethane-induced rats. Carcinogenesis. 1999;20:2267–72.

[84] Murillo G, Choi JK, Pan O, Constantinou AI, Mehta RG. Efficacy of garbanzo and soybean flour in suppression of aberrant crypt foci in the colons of CF-1 mice. Anticancer Res. 2004;24:3049–55.

[85] Feregrino-Perez AA, Berumen LC, Garcia-Alcocer G, Guevara-Gonzalez RG, Ramos-Gomez M, Reynoso-Camacho R, et al. Composition and chemopreventive effect of polysaccharides from common beans (*Phaseolus vulgaris* L.) on azoxymethane-induced colon cancer. J Agric Food Chem. 2008;56:8737–44.

[86] Faris MA, Takruri HR, Shomaf MS, Bustanji YK. Chemopreventive effect of raw and cooked lentils (*Lens culinaris* L) and soybeans (*Glycine max*) against azoxymethane-induced aberrant crypt foci. Nutr Res. 2009;29:355–62.

[87] Vergara-Castaneda HA, Guevara-Gonzalez RG, Ramos-Gomez M, Reynoso-Camacho R, Guzman-Maldonado H, Feregrino-Perez AA, et al. Non-digestible fraction of cooked bean (*Phaseolus vulgaris* L.) cultivar Bayo Madero suppresses colonic aberrant crypt foci in azoxymethane-induced rats. Food Funct. 2010;1:294–300.

[88] Feregrino-Perez AA, Pinol-Felis C, Gomez-Arbones X, Guevara-Gonzalez RG, Campos-Vega R, Acosta-Gallegos J, et al. A non-digestible fraction of the common bean (*Phaseolus vulgaris* L.) induces cell cycle arrest and apoptosis during early carcinogenesis. Plant Foods Hum Nutr. 2014;69:248–54.

[89] Boateng JA, Verghese M, Walker LT, Shackelford LA, Chawan CB. Inhibitory effects of selected dry beans (*Phaseolus spp* L) on azoxymethane-induced formation of aberrant crypt foci in Fisher 344 male rats. Nutr Res. 2007;27:640–6.

[90] Corpet DE, Pierre F. How good are rodent models of carcinogenesis in predicting efficacy in humans? A systematic review and meta-analysis of colon chemoprevention in rats, mice and men. Eur J Cancer. 2005;41:1911–22.

[91] Pegg AE. Methylation of the O6 position of guanine in DNA is the most likely initiating event in carcinogenesis by methylating agents. Cancer Invest. 1984;2:223–31.

[92] Clapper ML, Cooper HS, Chang WC. Dextran sulfate sodium-induced colitis-associated neoplasia: a promising model for the development of chemopreventive interventions. Acta Pharmacol Sin. 2007;28:1450–9.

[93] Kutos T, Golob T, Kac M, Plestenjak A. Dietary fibre content of dry and processed beans. Food Chem. 2003;80:231–5.

[94] Costa GED, Queiroz-Monici KDS, Reis SMPM, de Oliveira AC. Chemical composition, dietary fibre and resistant starch contents of raw and cooked pea, common bean, chickpea and lentil legumes. Food Chem. 2006;94:327–30.

[95] Guillon F, Champ MMJ. Carbohydrate fractions of legumes: uses in human nutrition and potential for health. Brit J Nutr. 2002;88:S293–S306.

[96] Olmedilla-Alonso B, Pedrosa MM, Cuadrado C, Brito M, Asensio-S-Manzanera C, Asensio-Vegas C. Composition of two Spanish common dry beans (*Phaseolus vulgaris*), "Almonga" and "Curruquilla", and their postprandial effect in type 2 diabetics. J Sci Food Agr. 2013;93:1076–82.

[97] Pedrosa MM, Cuadrado C, Burbano C, Muzquiz M, Cabellos B, Olmedilla-Alonso B, et al. Effects of industrial canning on the proximate composition, bioactive compounds contents and nutritional profile of two Spanish common dry beans (*Phaseolus vulgaris* L.). Food Chem. 2015;166:68–75.

[98] Bauer-Marinovic M, Florian S, Muller-Schmehl K, Glatt H, Jacobasch G. Dietary resistant starch type 3 prevents tumor induction by 1,2-dimethylhydrazine and alters proliferation, apoptosis and dedifferentiation in rat colon. Carcinogenesis. 2006;27:1849–59.

[99] Prado-Silva L, Azevedo L, Oliveira JAC, Moreira APM, Schmiele M, Chang YK, et al. Sesame and resistant starch reduce the colon carcinogenesis and oxidative stress in 1,2-dimethylhydrazine-induced cancer in Wistar rats. Food Res Int. 2014;62:609–17.

[100] Liu RP, Xu GF. Effects of resistant starch on colonic preneoplastic aberrant crypt foci in rats. Food Chem Toxicol. 2008;46:2672–9.

[101] Murphy MM, Douglass JS, Birkett A. Resistant starch intakes in the United States. J Am Diet Assoc. 2008;108:67–78.

[102] Le Leu RK, Brown IL, Hu Y, Morita T, Esterman A, Young GP. Effect of dietary resistant starch and protein on colonic fermentation and intestinal tumourigenesis in rats. Carcinogenesis. 2007;28:240–5.

[103] Pool-Zobel BL. Inulin-type fructans and reduction in colon cancer risk: review of experimental and human data. Brit J Nutr. 2005;93:S73–S90.

[104] Tosh SM, Yada S. Dietary fibres in pulse seeds and fractions: characterization, functional attributes, and applications. Food Res Int. 2010;43:450–60.

[105] Reddy BS. Role of dietary fiber in colon cancer: an overview. Am J Med. 1999;106:16S–9S; discussion 50S–1S.

[106] Coleman LJ, Landstrom EK, Royle PJ, Bird AR, McIntosh GH. A diet containing alpha-cellulose and fish oil reduces aberrant crypt foci formation and modulates other possible markers for colon cancer risk in azoxymethane-treated rats. J Nutr. 2002;132:2312–8.

[107] Wu WT, Yang LC, Chen HL. Effects of konjac glucomannan, inulin and cellulose on acute colonic responses to genotoxic azoxymethane. Food Chem. 2014;155:304–10.

[108] Raicht RF, Cohen BI, Fazzini EP, Sarwal AN, Takahashi M. Protective effect of plant sterols against chemically-induced colon tumors in rats. Cancer Res. 1980;40:403–5.

[109] Baskar AA, Ignacimuthu S, Paulraj GM, Al Numair KS. Chemopreventive potential of β-sitosterol in experimental colon cancer model -- an *in vitro* and *in vivo* study. BMC Complement Altern Med. 2010;10:24.

[110] Baskar AA, Al Numair KS, Paulraj MG, Alsaif MA, Al Muamar M, Ignacimuthu S. β-sitosterol prevents lipid peroxidation and improves antioxidant status and histoarchitecture in rats with 1,2-dimethylhydrazine-induced colon cancer. J Med Food. 2012;15:335–43.

[111] Bin Sayeed MS, Ameen SS. Beta-sitosterol: a promising but orphan nutraceutical to fight against cancer. Nutr Cancer. 2015;67:1214–20.

[112] Koratkar R, Rao AV. Effect of soya bean saponins on azoxymethane-induced preneoplastic lesions in the colon of mice. Nutr Cancer. 1997;27:206–9.

[113] Raju J, Patlolla JMR, Swamy MV, Rao CV. Diosgenin, a steroid saponin of *Trigonella foenum graecum* (fenugreek), inhibits azoxymethane-induced aberrant crypt foci formation in F344 rats and induces apoptosis in HT-29 human colon cancer cells. Cancer Epidemiol Biomarkers Prev. 2004;13:1392–8.

[114] Guajardo-Flores D, Serna-Saldivar SO, Gutierrez-Uribe JA. Evaluation of the antioxidant and antiproliferative activities of extracted saponins and flavonols from germinated black beans (*Phaseolus vulgaris* L.). Food Chem. 2013;141:1497–503.

[115] Puangpraphant S, Dia VP, de Mejia EG, Garcia G, Berhow MA, Wallig MA. Yerba mate tea and mate saponins prevented azoxymethane-induced inflammation of rat colon through suppression of NF-kB p65ser(311) signaling via IkB-α and GSK-3β reduced phosphorylation. Biofactors. 2013;39:430–40.

[116] Guo YW, Chen YH, Chiu WC, Liao H, Lin SH. Soy saponins meditate the progression of colon cancer in rats by inhibiting the activity of β-glucuronidase and the number of aberrant crypt foci but not cyclooxygenase-2 activity. ISRN Oncol. 2013;2013:645817.

[117] Champ MMJ. Non-nutrient bioactive substances of pulses. Brit J Nutr. 2002;88:S307–S19.

[118] Xu BJ, Chang SKC. Total phenolic, phenolic acid, anthocyanin, flavan-3-ol, and flavonol profiles and antioxidant properties of pinto and black beans (*Phaseolus vulgaris* L.) as affected by thermal processing. J Agric Food Chem. 2009;57:4754–64.

[119] Hayat I, Ahmad A, Masud T, Ahmed A, Bashir S. Nutritional and health perspectives of beans (*Phaseolus vulgaris* L.): an overview. Crit Rev Food Sci. 2014;54:580–92.

[120] Caprioli G, Giusti F, Ballini R, Sagratini G, Vila-Donat P, Vittori S, et al. Lipid nutritional value of legumes: evaluation of different extraction methods and determination of fatty acid composition. Food Chem. 2016;192:965–71.

[121] Billings PC, Newberne PM, Kennedy AR. Protease inhibitor suppression of colon and anal gland carcinogenesis induced by dimethylhydrazine. Carcinogenesis. 1990;11:1083–6.

[122] St Clair WH, Billings PC, Carew JA, Kellermcgandy C, Newberne P, Kennedy AR. Suppression of dimethylhydrazine-induced carcinogenesis in mice by dietary addition of the Bowman-Birk protease inhibitor. Cancer Res. 1990;50:580–6.

[123] Kennedy AR, Billings PC, Wan XS, Newberne PM. Effects of Bowman-Birk inhibitor on rat colon carcinogenesis. Nutr Cancer. 2002;43:174–86.

[124] Carli AD, Vieira PMD, Silva KTS, Cota RGD, Carneiro CM, Castro-Borges W, et al. Bowman-Birk inhibitors, proteasome peptidase activities and colorectal pre neoplasias induced by 1,2-dimethylhydrazine in Swiss mice. Food Chem Toxicol. 2012;50:1405–12.

[125] Clemente A, Arques Mdel C. Bowman-Birk inhibitors from legumes as colorectal chemopreventive agents. World J Gastroenterol. 2014;20:10305–15.

[126] De Mejia EG, Prisecaru VI. Lectins as bioactive plant proteins: a potential in cancer treatment. Crit Rev Food Sci. 2005;45:425–45.

[127] Muzquiz M, Varela A, Burbano C, Cuadrado C, Guillamon E, Pedrosa MM. Bioactive compounds in legumes: pronutritive and antinutritive actions. Implications for nutrition and health. Phytochem Rev. 2012;11:227–44.

[128] Liu B, Bian HJ, Bao JK. Plant lectins: potential antineoplastic drugs from bench to clinic. Cancer Lett. 2010;287:1–12.

[129] Obiro WC, Zhang T, Jiang B. The nutraceutical role of the *Phaseolus vulgaris* alpha-amylase inhibitor. Br J Nutr. 2008;100:1–12.

[130] Kune GA, Kune S, Watson LF. Dietary sodium and potassium intake and colorectal cancer risk. Nutr Cancer. 1989;12:351–9.

[131] Kennedy DA, Stern SJ, Moretti M, Matok I, Sarkar M, Nickel C, et al. Folate intake and the risk of colorectal cancer: a systematic review and meta-analysis. Cancer Epidemiol. 2011;35:2–10.

[132] Xu BJ, Chang SKC. Phytochemical profiles and health-promoting effects of cool-season food legumes as influenced by thermal processing. J Agric Food Chem. 2009;57:10718–31.

[133] Pretlow TP, Oriordan MA, Somich GA, Amini SB, Pretlow TG. Aberrant crypts correlate with tumor-incidence in F344 rats treated with azoxymethane and phytate. Carcinogenesis. 1992;13:1509–12.

[134] Shivapurkar N, Tang ZC, Frost A, Alabaster O. A rapid dual organ rat carcinogenesis bioassay for evaluating the chemoprevention of breast and colon cancer. Cancer Lett. 1996;100:169–79.

[135] Jenab M, Thompson LU. The influence of phytic acid in wheat bran on early biomarkers of colon carcinogenesis. Carcinogenesis. 1998;19:1087–92.

[136] Norazalina S, Norhaizan ME, Hairuszah I, Norashareena MS. Anticarcinogenic efficacy of phytic acid extracted from rice bran on azoxymethane-induced colon carcinogenesis in rats. Exp Toxicol Pathol. 2010;62:259–68.

[137] Prynne CJ, McCarron A, Wadsworth MEJ, Stephen AM. Dietary fibre and phytate - a balancing act: results from three time points in a British Birth Cohort. Brit J Nutr. 2010;103:274–80.

[138] Xu BJ, Chang SKC. A comparative study on phenolic profiles and antioxidant activities of legumes as affected by extraction solvents. J Food Sci. 2007;72:S159–S66.

[139] Zhang B, Deng Z, Tang Y, Chen PX, Liu R, Ramdath DD, et al. Effect of domestic cooking on carotenoids, tocopherols, fatty acids, phenolics, and antioxidant activities of lentils (*Lens culinaris*). J Agric Food Chem. 2014;62:12585–94.

[140] Bobe G, Wang B, Seeram NP, Nair MG, Bourquin LD. Dietary anthocyanin-rich tart cherry extract inhibits intestinal tumorigenesis in APC(Min) mice fed suboptimal levels of sulindac. J Agric Food Chem. 2006;54:9322–8.

[141] Cooke D, Schwarz M, Boocock D, Winterhalter P, Steward WP, Gescher AJ, et al. Effect of cyanidin-3-glucoside and an anthocyanin mixture from bilberry on adenoma development in the ApcMin mouse model of intestinal carcinogenesis--relationship with tissue anthocyanin levels. Int J Cancer. 2006;119:2213–20.

[142] Hao XP, Bose M, Lambert JD, Ju JY, Lu G, Lee MJ, et al. Inhibition of intestinal tumorigenesis in Apc(Min/+) mice by green tea polyphenols (polyphenon E) and individual catechins. Nutr Cancer. 2007;59:62–9.

[143] Bobe GY, M, Lanza E, Cross AJ, Colburn NH. Chemopreventive Effects of Kaempferol in Colorectal Tumorigenesis. In: Villers GF, Fougere Y, editors. Kaempferol – Chemistry, Natural Occurrences and Health Benefits. New York, NY: Nova Science Publishers, Inc.; 2013.

[144] Saud SM, Young MR, Jones-Hall YL, Ileva L, Evbuomwan MO, Wise J, et al. Chemopreventive activity of plant flavonoid isorhamnetin in colorectal cancer is mediated by oncogenic Src and beta-catenin. Cancer Res. 2013;73:5473–84.

[145] Gali-Muhtasib HU, Younes IH, Karchesy JJ, El-Sabban ME. Plant tannins inhibit the induction of aberrant crypt foci and colonic tumors by 1,2-dimethylhydrazine in mice. Nutr Cancer. 2001;39:108–16.

[146] Singletary KW, Meline B. Effect of grape seed proanthocyanidins on colon aberrant crypts and breast tumors in a rat dual-organ tumor model. Nutr Cancer. 2001;39:252–8.

[147] Nomoto H, Iigo M, Hamada H, Kojima S, Tsuda H. Chemoprevention of colorectal cancer by grape seed proanthocyanidin is accompanied by a decrease in proliferation and increase in apoptosis. Nutr Cancer. 2004;49:81–8.

[148] Kawabata K, Yamamoto T, Hara A, Shimizu M, Yamada Y, Matsunaga K, et al. Modifying effects of ferulic acid on azoxymethane-induced colon carcinogenesis in F344 rats. Cancer Lett. 2000;157:15–21.

[149] Wargovich MJ, Jimenez A, McKee K, Steele VE, Velasco M, Woods J, et al. Efficacy of potential chemopreventive agents on rat colon aberrant crypt formation and progression. Carcinogenesis. 2000;21:1149–55.

[150] Joseph OP, Phelomene M, Helene N, Valens H, Patrick OM, Thavarajah D, Thavarajah P. Phenolic compound profiles of two common beans consumed by Rwandans. Am J Plant Sci. 2014;5:2943–7.

[151] Balaji C, Muthukumaran J, Nalini N. Effect of sinapic acid on 1,2 dimethylhydrazine induced aberrant crypt foci, biotransforming bacterial enzymes and circulatory oxidative stress status in experimental rat colon carcinogenesis. Bratisl Med J. 2015;116:560–6.

[152] Thompson LU, Boucher BA, Liu Z, Cotterchio M, Kreiger N. Phytoestrogen content of foods consumed in Canada, including isoflavones, lignans, and coumestan. Nutr Cancer. 2006;54:184–201.

[153] Davis CD, Milner JA. Biomarkers for diet and cancer prevention research: potentials and challenges. Acta Pharmacol Sin. 2007;28:1262–73.

[154] Hanahan D, Weinberg RA. Hallmarks of cancer: the next generation. Cell. 2011;144:646–74.

[155] Daniell EL, Ryan EP, Brick MA, Thompson HJ. Dietary dry bean effects on hepatic expression of stress and toxicity-related genes in rats. Brit J Nutr. 2012;108:S37–45.

[156] Hayde VC, Ramon GG, Lorenzo GO, Dave OB, Rosalia RC, Paul W, et al. Non-digestible fraction of beans (*Phaseolus vulgaris* L.) modulates signalling pathway genes at an early stage of colon cancer in Sprague-Dawley rats. Br J Nutr. 2012;108 Suppl 1:S145–54.

[157] Campos-Vega RO, Dave Oomah D, Loarca-Pina G, Vergara-Castaneda HA. Common beans and their non-digestible fraction: cancer inhibitory activity—an overview. Foods. 2013;2:374–92.

[158] Kumar V, Sinha AK, Makkar HPS, de Boeck G, Becker K. Dietary roles of non-starch polysachharides in human nutrition: a review. Crit Rev Food Sci. 2012;52:899–935.

[159] Monk JM, Zhang CP, Wu WQ, Zarepoor L, Lu JT, Liu RH, et al. White and dark kidney beans reduce colonic mucosal damage and inflammation in response to dextran sodium sulfate. J Nutr Biochem. 2015;26:752–60.

[160] Gullon P, Gullon B, Tavaria F, Vasconcelos M, Gomes AM. In vitro fermentation of lupin seeds (*Lupinus albus*) and broad beans (*Vicia faba*): dynamic modulation of the intestinal microbiota and metabolomic output. Food Funct. 2015;6:3316–22.

[161] Schwabe RF, Wang TC. Bacteria deliver a genotoxic hit. Science. 2012;338:52–3.

[162] Zackular JP, Baxter NT, Iverson KD, Sadler WD, Petrosino JF, Chen GY, et al. The gut microbiome modulates colon tumorigenesis. MBio. 2013;4:e00692–13.

[163] Campos-Vega R, Garcia-Gasca T, Guevara-Gonzalez R, Ramos-Gomez M, Oomah BD, Loarca-Pina G. Human gut flora-fermented nondigestible fraction from cooked bean (*Phaseolus vulgaris* L.) modifies protein expression associated with apoptosis, cell cycle arrest, and proliferation in human adenocarcinoma colon cancer cells. J Agric Food Chem. 2012;60:12443–50.

[164] Cruz-Bravo RK, Guevara-Gonzalez RG, Ramos-Gomez M, Oomah BD, Wiersma P, Campos-Vega R, et al. The fermented non-digestible fraction of common bean (*Phaseolus vulgaris* L.) triggers cell cycle arrest and apoptosis in human colon adenocarcinoma cells. Genes Nutr. 2014;9:359.

[165] Campos-Vega R, Guevara-Gonzalez RG, Guevara-Olvera BL, Oomah BD, Loarca-Pina G. Bean (*Phaseolus vulgaris* L.) polysaccharides modulate gene expression in human colon cancer cells (HT-29). Food Res Int. 2010;43:1057–64.

[166] Thompson MD, Thompson HJ. Physiological effects of bean (*Phaseolus vulgaris* L.) consumption on cellular signaling in cancer. Cell Cycle. 2012;11:835–6.

[167] Thompson MD, Mensack MM, Jiang WQ, Zhu ZJ, Lewis MR, McGinley JN, et al. Cell signaling pathways associated with a reduction in mammary cancer burden by dietary common bean (*Phaseolus vulgaris* L.). Carcinogenesis. 2012;33:226–32.

[168] Heiden MGV, Cantley LC, Thompson CB. Understanding the Warburg effect: the metabolic requirements of cell proliferation. Science. 2009;324:1029–33.

[169] Levine AJ, Puzio-Kuter AM. The control of the metabolic switch in cancers by oncogenes and tumor suppressor genes. Science. 2010;330:1340–4.

[170] Howell JJ, Ricoult SJH, Ben-Sahra I, Manning BD. A growing role for mTOR in promoting anabolic metabolism. Biochem Soc T. 2013;41:906–12.

[171] Cruz-Bravo RK, Guevara-Gonzalez R, Ramos-Gomez M, Garcia-Gasca T, Campos-Vega R, Oomah BD, et al. Fermented nondigestible fraction from common bean (*Phaseolus vulgaris* L.) cultivar Negro 8025 modulates HT-29 cell behavior. J Food Sci. 2011;76:T41–7.

[172] Bobe G, Albert PS, Sansbury LB, Lanza E, Schatzkin A, Colburn NH, et al. Interleukin-6 as a potential indicator for prevention of high-risk adenoma recurrence by dietary flavonols in the polyp prevention trial. Cancer Prev Res (Phila). 2010;3:764–75.

[173] Mentor-Marcel RA, Bobe G, Barrett KG, Young MR, Albert PS, Bennink MR, et al. Inflammation-associated serum and colon markers as indicators of dietary attenuation of colon carcinogenesis in ob/ob mice. Cancer Prev Res (Phila). 2009;2:60–9.

[174] Moreno-Jimenez MR, Cervantes-Cardoza V, Gallegos-Infante JA, Gonzalez-Laredo RF, Estrella I, Garcia-Gasca TDJ, et al. Phenolic composition changes of processed common beans: their antioxidant and anti-inflammatory effects in intestinal cancer cells. Food Res Int. 2015;76:79–85.

[175] Celasco G, Moro L, Aiello C, Mangano K, Milasi A, Quattrocchi C, et al. Calcium butyrate: anti-inflammatory effect on experimental colitis in rats and antitumor properties. Biomed Rep. 2014;2:559–63.

[176] Arthur JC, Perez-Chanona E, Muhlbauer M, Tomkovich S, Uronis JM, Fan TJ, et al. Intestinal inflammation targets cancer-inducing activity of the microbiota. Science. 2012;338:120–3.